THE FRANKLIN D. MURPHY LECTURES X

THE FRANKLIN D. MURPHY LECTURES X

Art and the Natural World in Nineteenth-Century Britain: Three Essays

William Vaughan

SPENCER MUSEUM OF ART • THE UNIVERSITY OF KANSAS

MANAGING EDITORS: Catherine Evans and Sally Hayden
ASSISTANT EDITOR: Maggie Stenz and Ami Williams
EDITING: Annasue McCleave Wilson and Carol Shankel
GRAPHIC DESIGNER: Susan Hyde and Valerie Spicher

COLOR PLATE ON DUST JACKET: William Dyce
(Scotland, 1806–1864), *Pegwell Bay,* 1860, oil.
The Tate Gallery, London

ISBN: 0–913689–33–5

Contents

Preface

THESE THREE ESSAYS explore aspects of the relationship between the visual arts and scientific exploration of the natural world in the nineteenth century. I do not seek to give a comprehensive coverage of what is virtually an endless topic, but rather to pinpoint key issues. Central amongst these is the question of the status of naturalism and its relationship to religious belief and national identity. The areas covered by these essays have relevance for the history of European and American art in general during this period. I have restricted discussion largely to Britain because I feel that the issues can be explored in a more focused way if they are considered within the context of one particular culture.

The ideas behind these essays have been with me for some time. Chapter Two is a reworking of a lecture delivered at a symposium at the Zentralinstitut, Munich, in 1979, and subsequently published in the volume *Sind Briten hier?* by the Zentralinstitut and Wilhelm Fink Verlag. I am most grateful for the opportunity to reissue it here in an updated version. Chapter Three is in essence the lecture I delivered at the Spencer Museum of Art in November 1987. Chapter One relates to a lecture delivered the same week at the Nelson-Atkins Museum of Art, Kansas City, Missouri, and to the seminar course on nineteenth-century landscape I conducted at the University of Kansas.

I am extremely grateful to the University of Kansas, the Nelson-Atkins Museum of Art, and to the late Dr. Franklin Murphy for the opportunity to deliver these lectures and for the stimulus to bring them together in book form. Finally, I would like to thank Annasue McCleave Wilson and Sally Hayden for their most careful and thoughtful editing of my text and Susan Hyde and Valerie Spicher for making such a fine production out of it.

William Vaughan
Birkbeck College
University of London
April 1989

List of Illustrations

Pablo Picasso (Spain, 1881–1973)
3.1 *Les Desmoiselles d'Avignon*, 1907

Dante Gabriel Rossetti (England, 1828–1882)
3.8 *Ecce Ancilla Domini* [The Annunciation], 1850

John Ruskin (England, 1819–1900)
1.6 *Aiguille Drawing*

Margaret Stokes (Ireland, 1832–1900)
3.17 Title page to Samuel Ferguson's *The Cromlech on Howth*

Francisco Traini (Italy, 1321–1365)
3.7 Engraved by E. Lasinio, *The Last Judgment*

J. M. W. Turner (England, 1775–1851)
1.5 *The Mer de Glace, Chamonix*, 1802; 1.7 *The Fall of an Avalanche in the Grisons*, 1810; 1.8 *Snow Storm: Hannibal and his Army Crossing the Alps*, 1812

After Philipp Veit
2.2 *Christ Knocking on the Door of the Human Soul*, 1824

Horace Vernet (France, 1789–1863)
2.3 *Judith and Holofernes*, 1831; 2.4 After Horace Vernet, *Rebecca and Eliezer*, ca. 1840

Sir David Wilkie (Scotland, 1785–1841)
3.6 *Chelsea Pensioners Reading the News of the Battle of Waterloo*, 1819

Richard Wilson (England, 1714–1782)
1.4 *View of Holt Bridge on the River Dee*, 1762

1.12 *Dinner in the Iguanodon Model at the Crystal Palace on 31st December, 1853*; 3.2 *Apollo Belvedere*, ca. 350–320 B.C.; 3.3 "The Marquesas Islanders"; 3.4 "Native of the Sandwich Islands"; 3.5 "New Zealand War Canoes"; 3.9 "Indian Textile"; 3.10 "Decorated Maori Head"; 3.12 Example of "false" figurative art; 3.13 Example of "true" figurative art; 3.14 "Irish Manuscript"; 3.15 "Illuminated Manuscript, No. 2"; 3.16 *The High Cross of Monasterboice*; 3.18 "Celtic Ornament, No. 2"

I. "Those Dreadful Hammers"— Landscape, Religion, and Geology

SINCE THE TIME of the Renaissance the scientific investigation of nature has had direct implications for the painting of landscape. As knowledge about the natural world has changed so have people's expectations of its visual representation. Sometimes this is a relatively technical matter. As understanding of the formation of clouds developed, for example, the artist had to change his manner of representing them or run the risk of being accused of being "untrue" to nature. At other times the changes have been more profound and have affected not just the appearance of individual features, but the whole rationale behind a particular approach to landscape painting. The phenomenon described in this chapter falls into the latter category. For the assertions made by the new science of geology in the early nineteenth century had dramatic consequences for the pantheistic, romantic approach to landscape painting. Ultimately, in the Victorian period this was to be one of the principal factors in the retreat of this branch of the genre.

The critical issue at stake was the question of whether the natural world provided direct evidence of the existence of God—more specifically of the God described in the Old and New Testaments who was believed in, in a more or less literal sense, by most practicing Christians of the period. This issue was hardly a new one, but it took on a new urgency in the face of contemporary discoveries. In this context the new science of geology provided something of a test case as it was providing an unprecedented type of information about the natural world. The evidence provided by geology at that time seemed increasingly to run counter to the biblical account of creation. For most people this conflict is symbolized by the debate that culminated in the publication of Charles Darwin's *On the Origin of Species* in 1859. This book suggested that the different forms of life on earth had evolved by means of gradual "natural" processes rather than having been created in a few days by God in the manner described in Genesis. We now know that the conflict between Darwinian and biblical accounts was not as great as it first seemed. Certainly Darwin had not decisively rejected religious belief at the time of the publication of his book. But in the latter part of the nineteenth century positions hardened and people found themselves increasingly in the position of having to accept one account and reject the other.

The apparent threat to the literal reading of the Bible that the work of Darwin and his followers posed was ironic because the study of natural history before that time—at a popular level at least— had seemed to be providing strong reinforcement of proof of Divine existence. The famous "Bridgewater Treatises" written by William Buckland and other eminent natural scientists had been a seminal force in this direction for Darwin's own generation.[1] The ability of natural history to deliver proof of Divine order was indeed one of the fascinations of the subject for the Victorians. Editions of natural history books (we are told in works such as Lynn Barber's study of the subject[2]) often outsold popular novels. It helps to explain why the subject was so immensely popular.

Natural History, Religion, and Landscape

Over the last decade it has been recognized by many writers that British landscape painting during the period from 1830 to 1860 frequently exhibited responses to and interventions in the debate surrounding geological evidence and religious belief.[3] But the details of this history are still in the process of being gathered. It is perhaps for this reason that as yet there has not been much serious consideration of the implications of this debate for the practice of naturalistic landscape painting as a whole. It is this issue that will form the focus of the present chapter.

The religious view of natural history provided a unity of purpose between it and landscape painting of the period. The situation perhaps was made all the more striking because of the immense interest in and impressive performance of British landscapists of the period. It was, after all, an age in which British landscape painting was at its height because of figures such as Turner, Constable, Martin, and the great school of watercolorists.

Like the natural historians, these artists could see their exploration of the natural world as a demonstration of the existence of Divine power. The Romantic conception that the love of nature in itself was an agent of faith was constantly being reasserted. It still had currency in the 1850s, when John Ruskin—the greatest art critic of the age—stated in his book *Modern Painters*, "the love of nature, wherever it has existed, has been a faithful and sacred element of human feeling; that is to say, supposing all circumstances otherwise the same with respect to two individuals, the one who loves nature most will be *always* found to have more *faith in God* than the other."[4]

At the same time, however, artists were inclined to view their study of nature as providing a record comparable to that of a scientist. Constable, for example, went so far as to draw the connection specifically. In a lecture on landscape painting delivered toward the end of his life in 1836, he claimed: "Painting is a science, and should be pursued as an inquiry into the laws of nature. Why, then, may not landscape painting be considered as a branch of natural philosophy, of which pictures are but the experiments?"[5]

At this point the common interest in the laws of nature seems to have reinforced a line between scientist and artist. But it was a union that was not to last. By the end of the century there is a different picture, one in which artists have become aesthetes sharpening their poetic sensibilities and dismissing the world of science for its soulless materialism.

Science and Art in Earlier Centuries

Since what I am setting out to illustrate is the breaking down of a relationship, it is necessary for me to begin by describing what was breaking down. It is necessary to look, in fact, at the kind of interrelationships that existed in the early nineteenth century before the "fall" so to speak from unity of purpose.

As an art historian, I am hardly in a position to pronounce about the history of science. Fortunately, however, even the non-specialist is currently in a position to relay a lot of information about the development of the natural sciences in the early nineteenth century. Thanks to many excellent studies—such as Allen's *The Naturalist in Britain*, Roy Porter's *The Making of Geology*, and more recently Nicolaas Rupke's *The Great Chain of History*[6]—we now have a rich and detailed view of this field. While opinions between these experts differ, there has been broad agreement about the relationship of developments both to the need for greater knowledge of the natural resources being used by the expanding industries of the time and to the general aesthetic reevaluation of nature that followed Rousseau. "Never before or since," writes Allen, "has natural history been so much to the forefront in general aesthetic advance."[7] It is well acknowledged too, that natural history at this time was still at a stage of development to which the amateur could make an original contribution. One cause of this was the classification system set up in the mid-eighteenth century by Linnaeus. There were vast armies of amateurs, amongst whom clergymen were well to the fore,[8] who delighted in struggling up mountains with hammers and vascula charting the bold new frontiers of the natural world.

This popularity is important to remember in the context of my theme, for it means that there was a broad public who was well versed in scientific observation and the exploration of nature, and who was also keenly interested in the arts. The public's expectations of naturalism seem to have increased as a result. John Ruskin was an example of this new kind of amateur. One of the first things one notices in reading the first volume of his *Modern Painters* is his demonstration of intensive standards of observation. But Ruskin is at the end of a line that reaches back into the eighteenth century, to such figures as Erasmus Darwin, the patron of Joseph Wright of Derby, or, to take an example from a later period, to Dawson Turner, the naturalist and patron of the Norwich landscapist John Crome.

Of course this period is far from being the first time in which a great naturalistic advance in art has run parallel with rapid developments in the natural sciences. Italy in the fifteenth century and Holland in the seventeenth century offer similar examples. These earlier examples can serve to remind us how much such movements lead to a change in the perception of the natural world. For in each case the concept of what looked "naturalistic" altered dramatically. It is significant that the early nineteenth century is the time when the naturalism of Dutch seventeenth-century artists—while still admired—was no longer seen as providing a comprehensive model. After a century in which landscape had tended to concentrate more upon stylization and the elaboration of well worked out formulas, in which harmony and arcadian ease were set at a higher premium than realism, we find once more a reversal of roles. Exploration and discovery, once again, have supremacy over the creation of an idyll.

It is undoubtedly the case that the growth of interest in the natural sciences at the end of the eighteenth century heightened the impetus towards naturalism in landscape painting. But there was a theological impetus as well. This was the return—through Romanticism—of the idea of perceiving God directly in nature. While this feature was pan-European—as evident, for example in the works of Rousseau in France, Goethe in Germany, and Wordsworth in England—it is worth remembering that this link had a particularly strong presence in the natural sciences in England. In *The Great Chain of History*, Nicolaas Rupke has drawn attention to the distinct nature of the "English" school of geology that flourished in the 1820s and 1830s. This consisted largely of the Oxford scholar William Buckland and his followers. The earlier continental development of geology had drawn upon a historical framework established by the philosophy of history of Kant and Hegel. In England, where such an influence was resisted, there was no such model to hand. There was instead a far greater reliance on the model of early history provided in the Bible. The fact that so many of the English geologists—including Buckland—were Anglican clergymen helped to strengthen this dependence. Another feature of the English school that gave urgency to the issue of the relation of faith to scientific discovery was the particular fascination in England with paleontology, in the evidence provided by fossils of extinct species. The predominance of secondary rocks in England—those strata in which such finds were most likely to occur—made such a tendency even more pronounced. Since such species presented the area in which the new science was most obviously at variance with biblical history, there was all the more need to attempt a reconciliation. In his characterization of the English school of geology, Rupke makes a distinction not only with that on the continent—particularly in Germany—but also with Scotland. Here the approach adopted by Hutton and Playfair was far closer to the continental model in its reliance on philosophy. Furthermore, there was a tendency in Scotland to concentrate on mineralogy rather than paleontology (partly encouraged by the greater presence of "primary" rocks, which by their very nature do not contain fossil remains). This encouraged the Scottish geologists to concentrate more on the extent of time involved in early history rather than to become engrossed in the detailed relation of the stages of the earth's history. As an extension of this issue, they tended to regard the tortuous attempts to reconcile geological evidence and the biblical account of the early history of the world as being relatively trivial and beside the main point.[9] This distinction between English and Scottish geological practice at the time, I hope to show, did have some bearing upon the representation of geological features in landscapes by British artists. It is also striking that while the effects of recent geological discoveries can be seen in the landscapes of many continental artists of the period, none of them reveal the concern for the relationship between biblical narrative and the earth's history to be found in the work of British artists of the period. The "high mountain" paintings of the Swiss artist Caspar Wolff and of the German Romantic painters such as Joseph Anton Koch, Caspar David Friedrich, and Carl Gustav Carus incorporate much recently discovered information about rocks. But such information seems to have been used mainly to enhance the exactness of naturalistic description rather than to lead to any reflections on faith and biblical history.[10] A similar development can be found in America, where geological exploration encouraged a positively ecstatic view of landscape.[11]

It is now time to turn from the general to the particular. While the concern for geology and its finds was widespread amongst British landscapists of the early nineteenth century, their responses were significantly different. I have therefore decided to continue by looking at individual cases, through which it is hoped some sense of the complexity of the situation will be gathered, as well as a sense of a gradual shift of concerns. The artists I will be considering are John Constable, J. M. W. Turner, John Martin, the Pre-Raphaelites, and William Dyce.

Constable

CONSTABLE IS AN ARTIST who has gone down in history as a martyr for naturalism. From the time—as a young man of twenty-six in 1802—when he declared, "there is room enough for a natural painture,"[12] he appears to have dedicated himself to the task of representing his local Suffolk countryside as it was, without bowing to convention. It took decades before his uncompromising vision gained acceptance.

Constable's love of "factual" landscape was certainly stimulated by his background. The son of a wealthy landed miller, he was used to a practical view of the countryside, in which the terrain was explored for its productive potential. Like so many of his class—on the fringes of the gentry—he was fascinated by

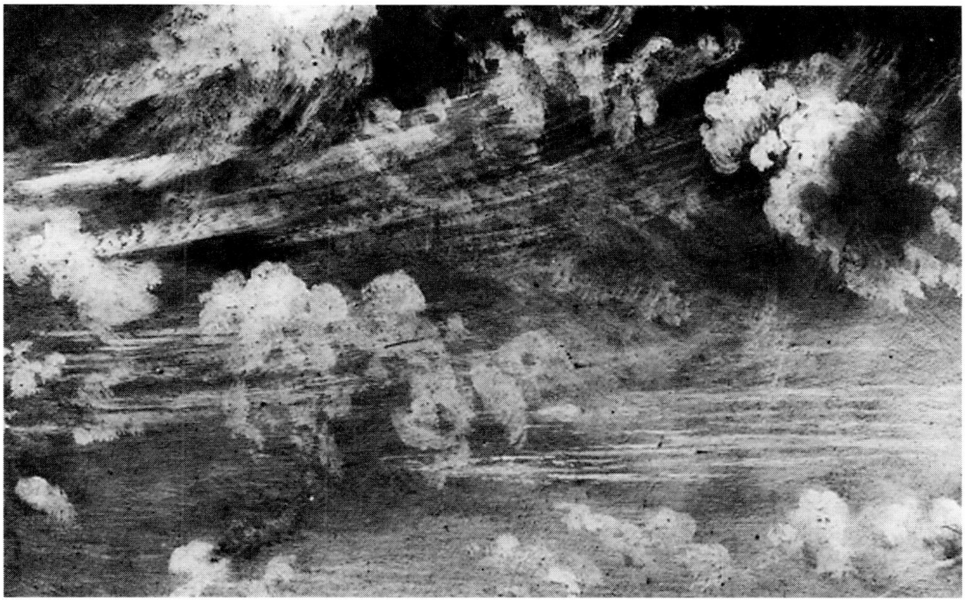

Figure 1.1. John Constable (England, 1776–1837), *Study of Cirrus Clouds*, ca. 1822, oil. Victoria and Albert Museum, London

scientific explorations of natural phenomena for both practical and aesthetic reasons. Toward the end of his life he professed to prefer science books to all other kinds of reading, and amongst science books those he liked best were on geology: "I must say that the sister arts have less hold on my mind in its occasional ramblings from my one pursuit than the sciences, especially the study of geology, which, more than any other, seems to satisfy my mind."[13]

I have already referred to his claim that pictures should be seen as "experiments" into the laws of nature. Certainly he took his desire to experiment a long way. In his studies he analyzed different types of natural phenomena. Spurred on by the fledgling science of meteorology he made oil sketches of different types of clouds, sometimes employing the nomenclature recently introduced by the pioneer meteorologist Luke Howard.[14] The inscription of the word "Cirrus" by the artist on the back of the picture reproduced in figure 1.1 is a particular instance of such usage.

It has often been objected that Constable's pictures—despite all this—are far from being raw records of natural observation. It has been shown that many of his Suffolk landscapes derive their compositions from the classical landscape painters of the seventeenth century.[15] But to call Constable "unnatural" on this account is to miss the point. In the first place, Constable saw himself as carrying on and updating the exploration of nature begun by such illustrious forebears as Claude. Secondly—and more importantly—he was concerned about the role played by structure in the elucidation of observations. Like so many of the natural scientists of the period he saw the "structure" in which nature became comprehensible that which demonstrated the natural order and harmony of the universe. Such an assumption had been a common feature in rationalist thought since the seventeenth century. In their monograph on Constable, Leslie Parris and Conal Shields have argued that the artist followed in particular the great seventeenth-century scientist Francis Bacon in believing that a multitude of separate observations—in this case his painted studies from nature—would inform and reveal this larger structure.[16]

It was this desire to make evident the natural order that encouraged

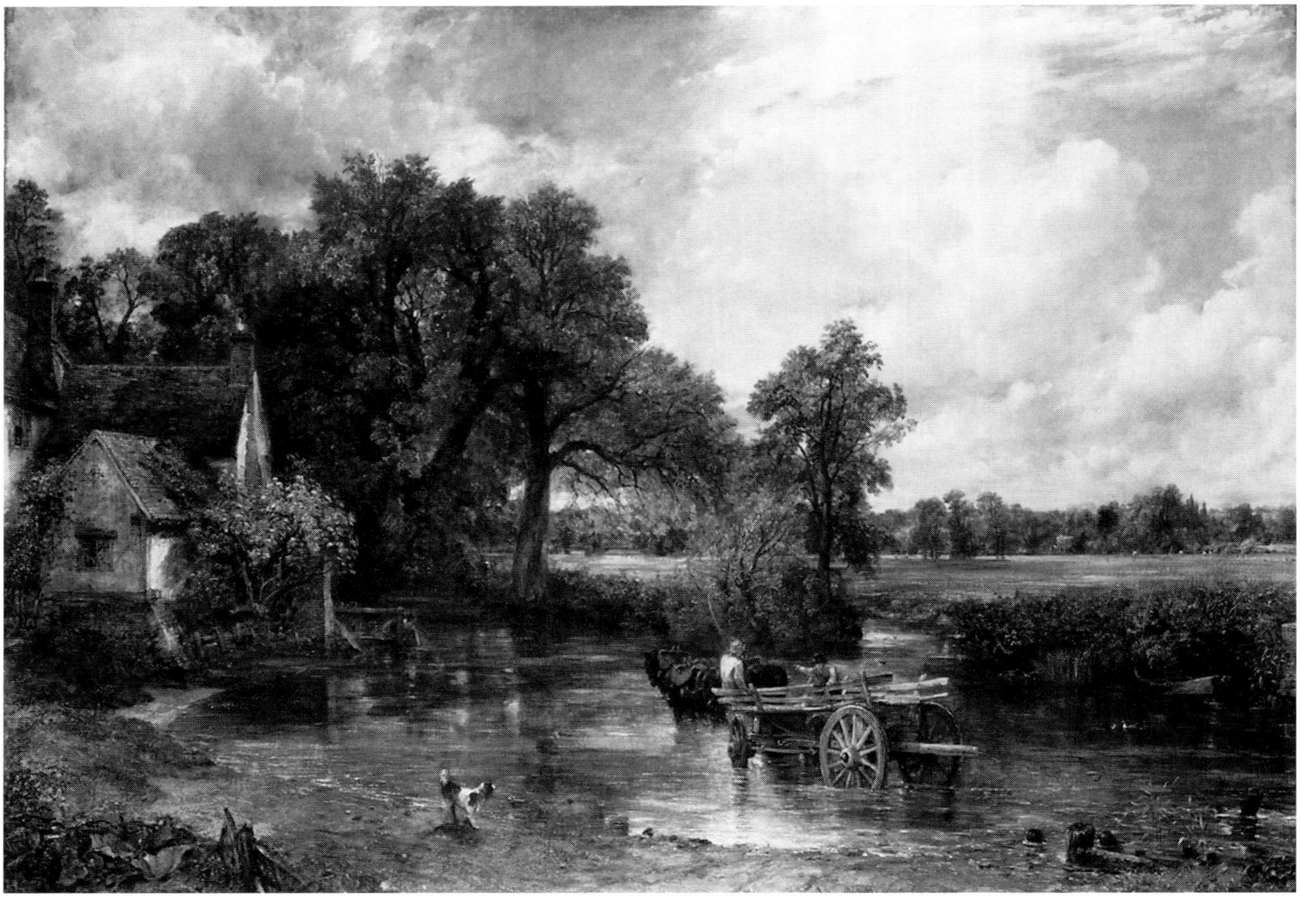

Figure 1.2. John Constable (England, 1776–1837), *The Haywain*, 1821, oil. National Gallery, London

Constable to construct out of his delightful small studies of corners of Suffolk such grand orchestrations as *The Haywain* (fig. 1.2). As well as the evident change of scale and revision of the internal forms to provide more "harmonious" intervals, it is also significant that in the later stages of the work man enters the scene in the form of the haywain and its driver. Man is still at the center of Constable's world; it is for him that it was created and it was he who exploited it most productively. The rich farming lands of Suffolk seem a natural place to make such a claim, especially from someone whose family owned a share of it. Significantly, Constable was far less enthusiastic about barren mountain scenery, where man's relationship with the elements was more questionable.

This interest in the exploration and demonstration of natural order gave an affinity between scientist and artist that provided Constable with much satisfaction. This is seen in a celebrated passage on an earlier British landscape painter whom Constable—along with so many of his generation—venerated: Richard Wilson. Constable envisaged Wilson in Paradise, walking arm in arm with Linnaeus and Milton.[17] The choice of figures is revealing. Milton was the great poet of English classicism who brought a sense of measure and order to his lyrical descriptions of the countryside. Wilson had imposed classical order on the British countryside. His *View of Holt Bridge on the River Dee* (fig. 1.4), for example, gives this celebrated meeting place between England and Wales a Claudian aspect. Wilson's other celestial companion, Linnaeus, (as has already been mentioned) was the great scientist of natural order whose binomial system of classification provided the key for the systematic organization of species.[18] Indeed, it has been claimed that Linnaeus's system was the single greatest factor in bringing about the vast popularity of natural history around 1800. Armed with his clear binomial system every amateur could rush to the field in the hope of identifying and adding to the list of known species. Indeed, there was a virtual compulsion to do so. For until the job of classification had been completed, the understanding of Divine creative intention, God's "great plan" as Linnaeus put it, would be incomplete.[19]

Both artist and scientist were engaged in perceiving the "great plan." But there was a difference of emphasis. Whereas the scientist's principal task was analysis, that of the artist was revelation. This brings one back to the exciting sense of harmony in Constable's great works—which is perhaps why he remains popular to this day. He really does bring alive a feeling of things connecting. And this is made the more poignant because the discovery of harmony was, for him, achieved only after a struggle. He was surprised at the "novel look" of *The Haywain* when he finished it; he had not calculated that it would work out that way.[20]

It is also clear—perhaps because strengthened by the sense of the fortuitous element in the process—that this "harmony" had a religious dimension for Constable. Although not given to elaborate symbolism he did feel his portrayal of nature was underpinned by an intuition of the Divine. Once when out walk-

Figure 1.3. John Constable (England, 1776–1837), *Willy Lot's Cottage*, 1814, oil. Victoria and Albert Museum, London

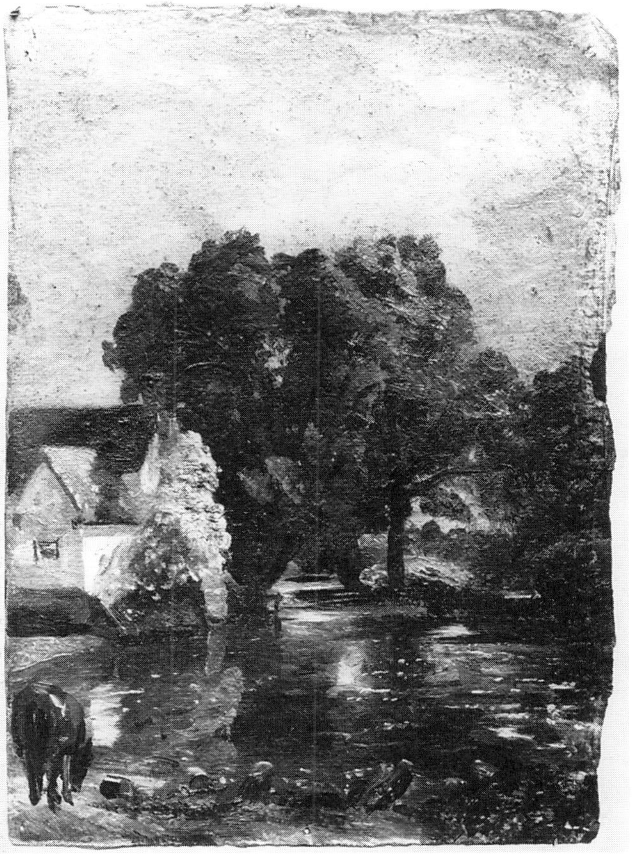

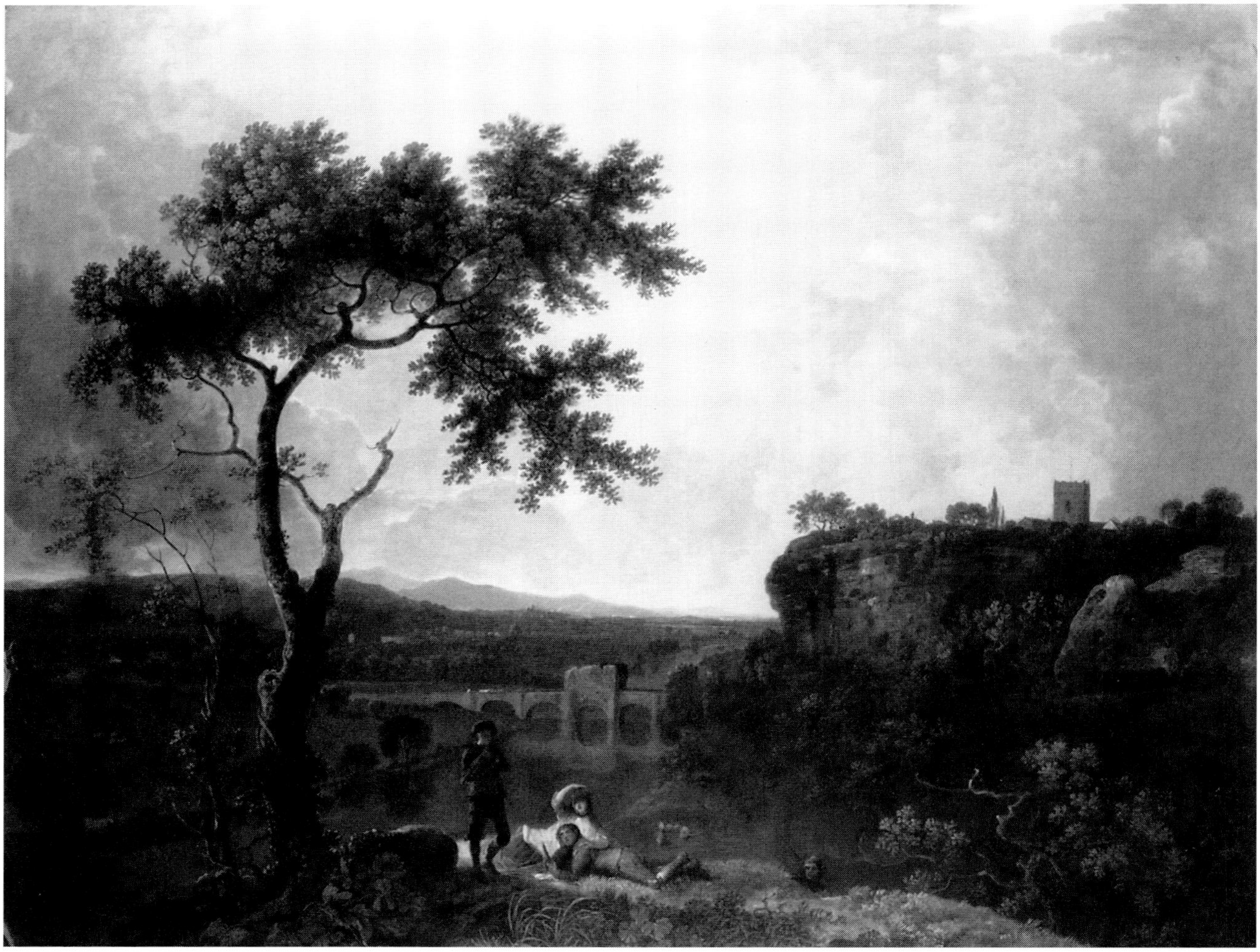

Figure 1.4. Richard Wilson (England, 1714–1782), *View of Holt Bridge on the River Dee*, 1762, oil. National Gallery, London

ing with Wordsworth in the spring he commented how the words "I am the resurrection and the life" seemed to be exemplified by the burgeoning forms around them. Constable was a churchman and ecclesiastics—most notably Archdeacon Fisher—were amongst his friends and patrons. His religion was of the traditional, unfanatical "broad church" kind, very similar, in fact, to that of Buckland and other leaders of the "English" school of geology. The fact that it was "broad" and rarely articulated did not mean that it was any the less serious for that.

Unfortunately for Constable the security and harmony conveyed by both social and aesthetic "structure" became less clear to him in later years. This can be seen in such works as his *Salisbury Cathedral* of 1831 where the bastions of natural order in the English countryside—the stately trees and the spire of the Anglican cathedral—are sorely pressed by the elements. There is, it is true, a rainbow of promise in the sky, but it is a close run thing, and should at the very least be taken as a warning. Biographical reasons for this darkening vision have frequently been cited: the grief caused by the death of his beloved wife, the mortification at the continued indifference of the artistic establishment, and his fears for the undermining of the social order through legislative changes such as the bill that emancipated Roman Catholics in 1829 and the Reform Bill of 1832.

Doubtless all these played their part. But Constable was an intelligent and reflective man; he would not have allowed such feelings to become evident unless they could have contributed to some system of thought. There were intellectual as well as emotional and social reasons for the disquiet of these later works.

It is clear that Constable was shifting his reading at this time more and more toward the study of the sciences. It has already been mentioned how he expressed a particular preference for reading geology books in the last years of his life, that is in the 1830s. Such a preference was exceedingly modish. For this new science—with its enthralling vision of the history of the earth, its uncovering of monstrous animal forms and indications of unimaginable eons of time—was attracting increasing popular interest at the time. It would be interesting to know what Constable actually had read, for the popular books of the period were divided on their view of whether the evidence provided by geology went against or supported a literal reading of the book of Genesis. If Constable had read Charles Lyell's *Principles of Geology*,[21] the most popular of these studies, he would have encountered a popularization of Hutton's concept of "uniformitarianism," the image of change taking place over vast eons of time by means of gradual processes still in operation. However, there is no evidence of the artist having done so; it was not amongst the contents of his library.[22] One can be fairly certain, on the other hand, that he was familiar with the more reassuring work of the equally popular French scientist Baron Cuvier, for that writer's *The Animal Kingdom* (translated in 1834–37 by G. Henderson) was in the his library. Cuvier's theories might have had a particular relevance for Constable in his later years. For, as shall be seen later, Cuvier—an opponent of "uniformitarianism"—described the history of development on earth in terms of a series of gigantic natural "catastrophes." There is certainly a sense of impending natural doom in Constable's elemental later works. In such pictures as *Salisbury Cathedral* (1831) and *Hadleigh Castle* (1829) nature appears to be engaged in struggle and conflict.

Figure 1.5. J. M. W. Turner (England, 1775–1851), *The Mer de Glace, Chamonix*, 1802, pencil and watercolor with body color on a prepared gray ground. The Tate Gallery, London

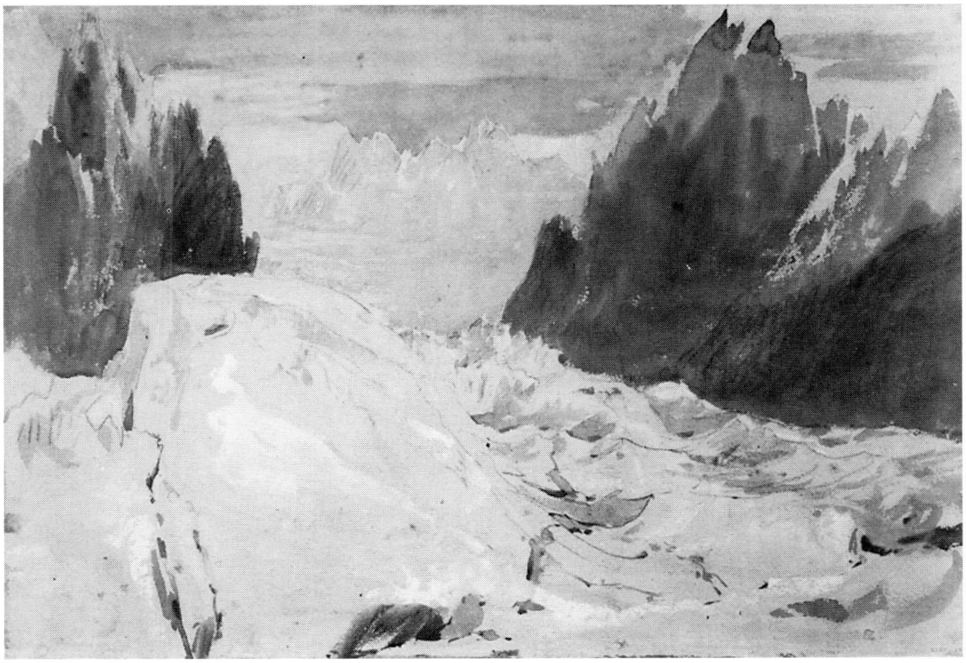

Perhaps Constable could see in the social, religious, and political turmoil of his own age a necessary stage of crisis in the unfolding of a "great plan." Cuvier's account would have assured him that the "physico-theological" basis of the Baconian world view need not be abandoned—only revised. It is striking that for all his personal pessimism Constable shows the sturdy symbols of the old world—the cathedrals, castles, and monuments—as enduring transient storms. If this is the case, perhaps it was fortunate for him that he died in 1837, before the greater turmoil of the Victorian world arrived.

Turner

THE EXTENT TO WHICH Constable's professed passion for geology affected the vision of his later pictures must remain a matter for speculation. But there were other artists of his generation who most explicitly employed the theories and observations of the new science.

In the nature of things, these artists tended to be those who chose a wider range of subject matter, in particular those who were servicing the new vogue for travel by producing pictures of far-off places and wild mountainous regions where the structure of rocks and violent earth movements could be most readily discerned. For, as Roy Porter has pointed out, the late eighteenth-century penchant for travel was one of the key stimuli that led to the development of geology as an independent practice.[23]

Paramount amongst the artists who explored this taste for travel to wild places was J. M. W. Turner. It is symptomatic of his interests that when he first had the opportunity to go abroad—as a young man of twenty-seven in 1802—he should have made straight for the Alps. His fascination for earth sciences is made clear in the watercolor that he made on this trip of the glacier at Chamonix (fig. 1.5). This is not simply because the picture contains a glacier and some striking rock formations, but also because it shows the hut erected in

Figure 1.6. John Ruskin (England, 1819–1900), *Aiguille Drawing*, date unknown, engraving, from Ruskin, *Modern Painters*, vol. IV, plate 32 (London, 1856).

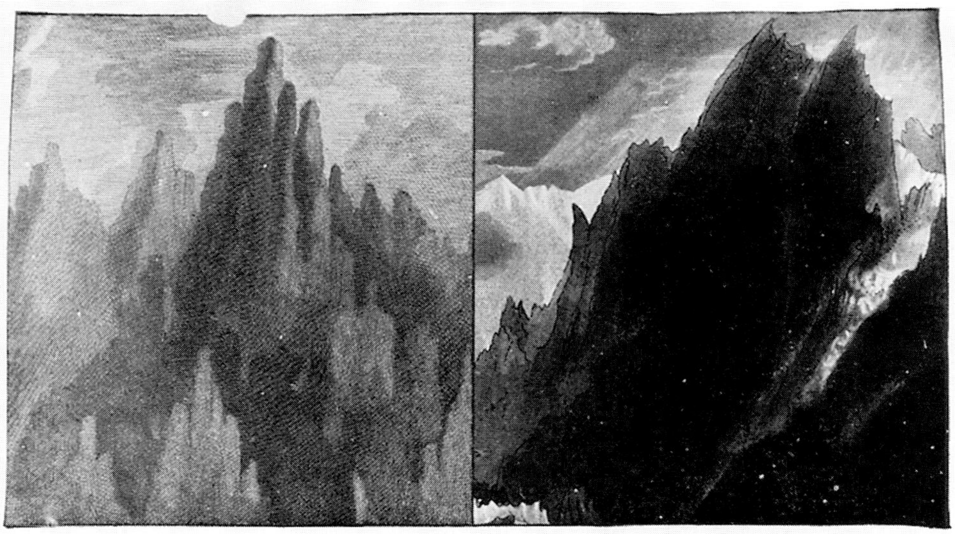

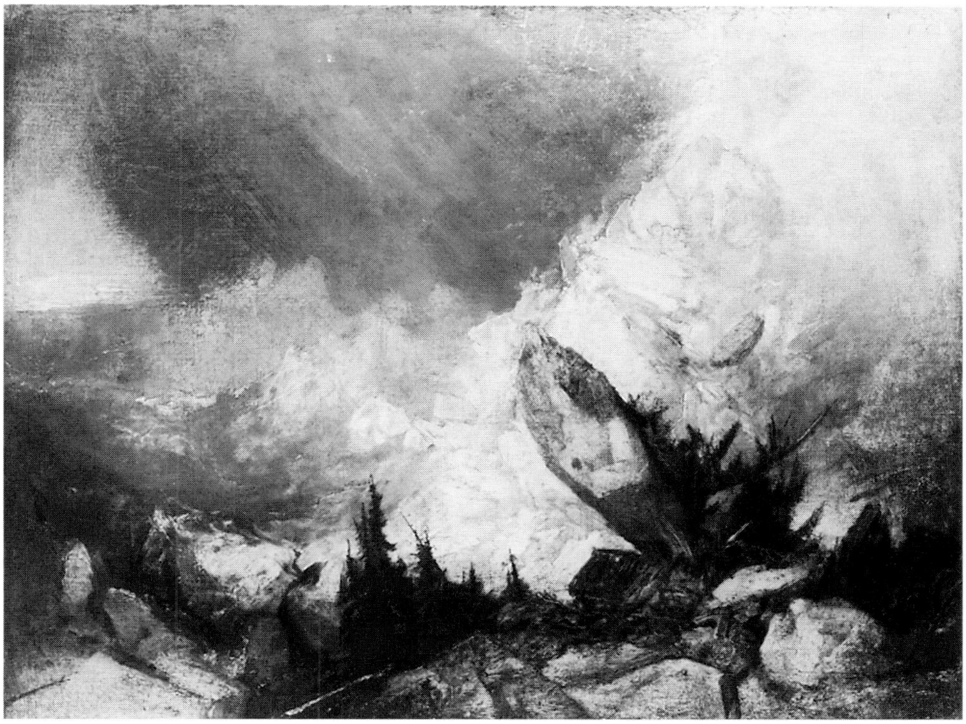

Figure 1.7. J. M. W. Turner (England, 1775–1851), *The Fall of an Avalanche in the Grisons*, 1810, oil. The Tate Gallery, London

1779 by Charles Blair, the celebrated British alpinist who had made a remarkable series of observations of the terrain a generation before.

We know much about Turner's interest in the science of optics, in perspective, and color theory.[24] We know less about the extent of his interest in and understanding of geology. However, there is evidence to suggest that he took an active interest in it during the first decades of the nineteenth century—the period when he was particularly active painting Alpine and high mountain subjects. He possessed the first two volumes of the *Transactions of the Geological Society* (published respectively in 1811 and 1814) and was a friend of John McCulloch, the first president of the society.[25] And there was a commentator of the period who was convinced that Turner's understanding of earth science—if not his study of its literature—was paramount. This was his great defender John Ruskin.

Ruskin, as has already been mentioned, was an enthusiastic geologist. In his autobiography, *Praeterita*, he claimed to have studied the subject from infancy. Indeed, his first introduction to the Alps had been through this medium, since he had read carefully the Swiss naturalist H. Benedict de Saussure's *Voyages dans les Alps* (1780–96)[26] before visiting the region. When Ruskin finally did encounter the mountains in real life, he found he had not been misled by his mentor. The effect was so striking to him that "the seen walls of lost Eden could not have been more beautiful to us."[27]

Ruskin's view of the Alps is a striking example of how, at this stage, the aesthetic and the scientific could be intermingled. Indeed, de Saussure had encouraged this, since he made a special issue of the virtue of close observation. This was a creed to which Ruskin willingly subscribed. And it gave him a chance to suggest that the perceptive eye of the artist could match—indeed even surpass—that of the field scientist in describing the structure of natural phenomena. In Volume Four of *Modern Painters* Ruskin used an alpine picture painted by

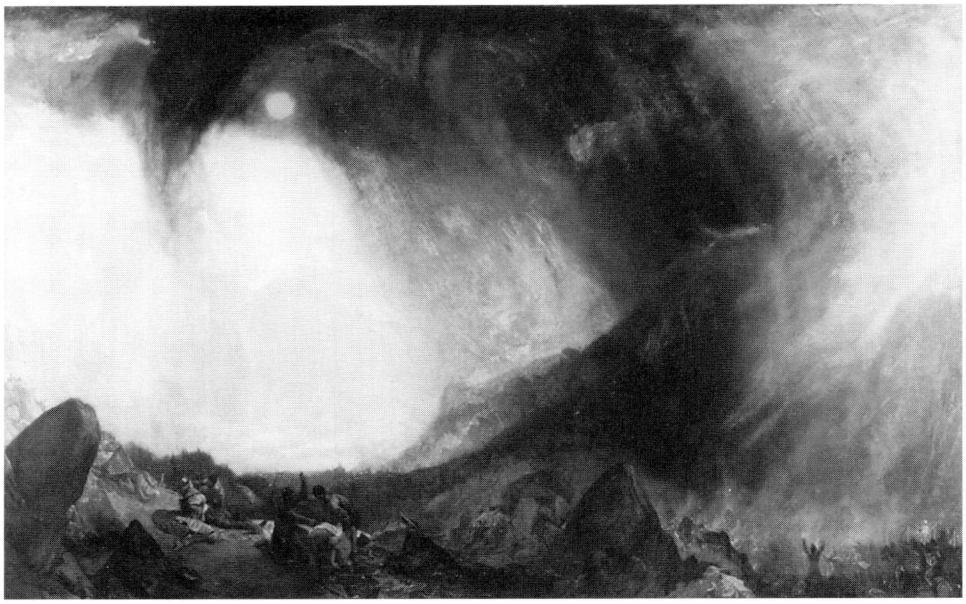

Figure 1.8. J. M. W. Turner (England, 1775–1851), *Snow Storm: Hannibal and his Army Crossing the Alps*, 1812, oil. The Tate Gallery, London

Turner to demonstrate his point (fig. 1.6). In the text he remarks how he has, after ten years of observation, finally managed to unravel the structure of the curved aiguilles of rock above the Chamonix glacier, showing that the sharp edges are not points, but plates of "sinewy granite." He is proud of his observation. "Even of professional geologists" he claims, "the only person who has described it properly is de Saussure whose *continual* sojourn amongst the Alps enabled him justly to discern the constant from the inconstant phenomena."

But there is more than personal boasting here. For Ruskin goes on to make clear that his achievement is nothing compared to that of his hero: "And yet in his very first journey to Savoy, Turner saw it at a glance, and fastened on it as the main thing to be expressed in those mountains."[28]

Ruskin's drawing purports to be a sketch of Turner's discovery. He may, one feels, have been leading his witness. But it is certainly clear that Turner was bringing a new level of acuity to the depiction of Alpine forms. In his large oils Turner used this knowledge to dramatic effect—notably in *Avalanche in the Grisons* (fig. 1.7) and *Hannibal Crossing the Alps* (fig. 1.8).

For Ruskin there was a theological point to be gained from this emphasis on acute observation of nature; aesthetic delight in mountains was a sensuous confirmation of the spiritual truth of the geologist's findings. Talking again of his emotion at seeing the Alps for the first time he said, "The first sight of the Alps had been to me a direct revelation of the benevolent will in creation." He then went on to make the sweeping claim already alluded to that "the one who loves nature most will be *always* found to have *more faith in God*."[29]

Ruskin may have been right in his assertion of Turner's acute insight into rock formations. But he seems to be less near the truth when he associates this acuity with the intimation of Divine benevolence. Turner seems to have had no such concern. It is the indifference of the universe to man's fate, by contrast, that seems to be his principal theme. His Alpine subjects dwell on danger and destruction, on the indifference of natural forces to man. In the *Avalanche* a humble cottage is being smashed to pieces. In *Snow Storm* the herculean task by the great Carthaginian general Hannibal of leading elephants across the Alps is sabotaged by skirmishes with tribesmen and a snowstorm. Worse than that (as Turner's accompanying verses make clear), the whole project is fruitless—for when Hannibal finally struggles through to Rome, he is defeated.[30]

In fact Turner's view of the elements is a pessimistic one—a view summed up by his long, unfinished poem *Fallacies of Hope*. There is certainly a God in Turner's cosmology, but he is apparently indifferent to human fate. Armed with such a view, Turner can have had none of the difficulties with the findings of the geologists that Ruskin and so many other enthusiasts for the subject were later to experience.

John Martin

As it happens, Ruskin's religious view of geology accords more closely to that of an artist for whom he had little sympathy—the popular landscape fantasist John Martin.

Martin arrived in London from Newcastle as a young man of twenty in 1809.[31] At that time Turner was undoubtedly the most dominant of dramatic landscape painters. Martin set out to challenge his position. In his early years he often borrowed features directly from Turner's work. His *Joshua Commanding the Sun to Stand Still* (fig. 1.9) shows a dependence upon *Hannibal Crossing the Alps* (fig. 1.8), particularly in the broad sweep of the stormy heavens. But there are two main differences—one of presentation and one of interpretation. On the technical side, Martin has replaced Turner's breadth of treatment with an elaboration of detail. His fantasy is at the same time both more extravagant and more literal than Turner's. The mountains and figures can be counted. Indeed, dramatic (yet precise) calculation was one of the attractions upon which Martin relied. In the pamphlets he frequently published to accompany his pictures he was in the habit of giving the exact height of his buildings and mountains. The calculated enormity had a similar appeal to that of science fiction today. It was an attitude that—as will be seen—encouraged links between his work and that of popular scientists of the time.

The difference in interpretation was also relevant to this relationship. In contrast to Turner, there is nothing ironic or subversive about Martin's work. His *Joshua* reinforced traditional religious faith. It showed a triumphant moment from the Old Testament when God made the sun stand still in the heavens to allow the Israelites a little extra time in which to smite their enemies. While Turner's *Hannibal* depicts a hero at odds with the universe, Martin's *Joshua* shows a faithful servant of the Lord being rewarded with a little bending of the laws of nature. The natural world in Martin's picture is on the side of the righteous and actively obeys the will of God.

Martin appealed to a more popular taste than Turner both in his calculated extravagance and in his affirmation of the Old Testament image of God in nature. He also shared the popular fascination with science and technology. He designed many engineering projects, ranging from a proposal for much-needed improvements to London's waterworks to the more questionable invention of an "elastic" iron ship intended to bend before the onslaught of the waves and thereby avoid damage. It was symptomatic of such interests that he took care to keep his pictures in line with new developments. Since he was a landscape painter, this involved a knowledge of geology, which he often used to overpowering effect.

One such triumphant usage was that in the scene of the *Deluge* that he painted in 1826. This work is now known only through the mezzotint the artist

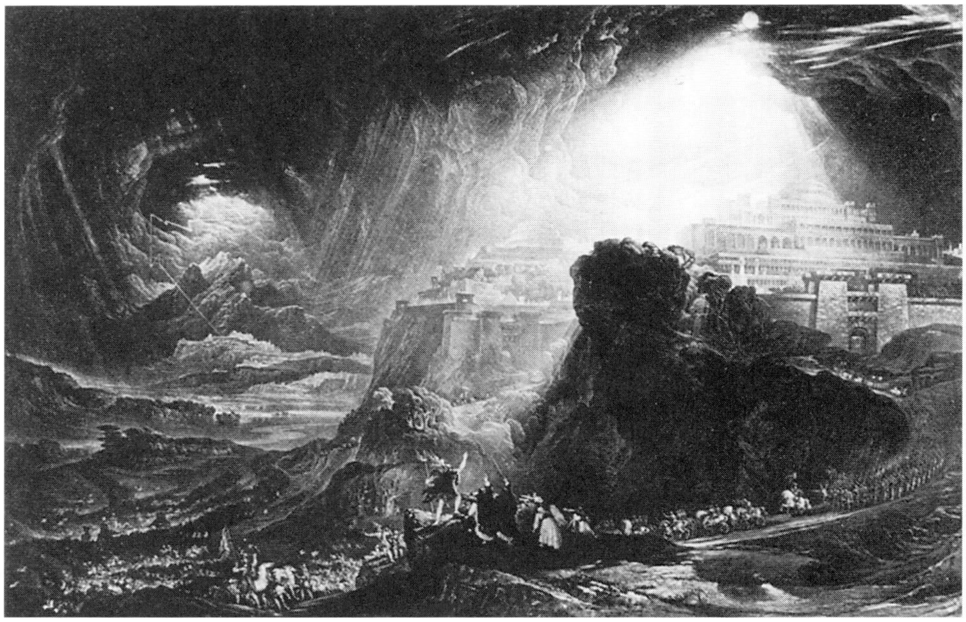

Figure 1.9. John Martin (England, 1789–1854), *Joshua Commanding the Sun to Stand Still upon Gideon*, 1816, oil. United Grand Lodge of Great Britain

Figure 1.10. John Martin (England, 1789–1854), *The Deluge*, 1828, mezzotint engraving. British Museum, London

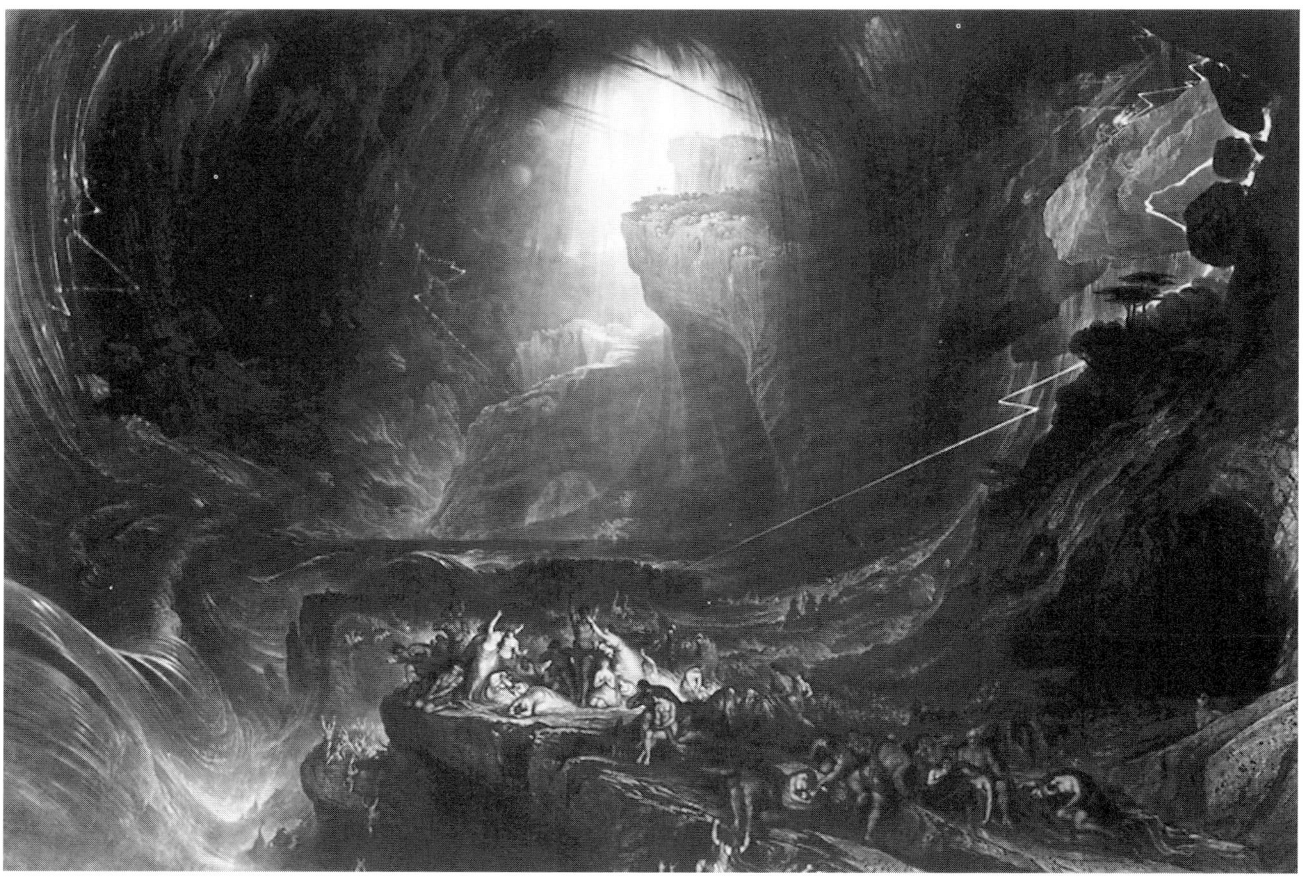

made of it two years later (fig. 1.10)—though a later version of the subject exists in the Yale Center for British Art, New Haven.[32] The great biblical flood had long been an issue of concern for geologists. The prevalence of fundamentalist beliefs gave an urgency to the quest for narratives that could reconcile recent discoveries with the account of creation in Genesis. During the 1820s such a narrative was provided with great success by the French geologist and paleontologist Baron Cuvier—the man whose works were read by Constable. Cuvier's *Recherches sur les Ossemens fossiles* (1811)[33] provided an attractive explanation of the puzzling fact that fossilized remains could be found of animals not mentioned in the Bible. He proposed that the early history of the earth had been marked by a series of "catastrophes," each of which left some of the species created by God in the first week of the earth's history extinct. It is important to note here that such a theory got around the need to subscribe to the concept of evolution, that is, that animals have gradually developed towards their present form rather than having been created in immutable shape. We may have lost some species since the Creation, Cuvier is saying, but we haven't made any more. The biblical deluge was, in Cuvier's opinion, the last of these destructive catastrophes.

Cuvier's concept was a brilliant projection. It brought together the Romantic love of sublime horror with the disturbing new evidence provided by paleontology and reconciled these with orthodox religion. Hardly surprising, therefore, that the idea went down so well at a popular level, or that Martin—a man profoundly religious, scientific, and desirous of acclaim—should have followed his lead. In Martin's own account of the Deluge, some highly detailed references to Cuvier have been made.[34] The artist believed the French scientist, for example, in supposing that the violent upsurge of water was caused by the sun, moon, and a comet coming into a conjunction and drawing the water over the earth.[35] Another borrowing can be seen in the way Martin shows the "lost" prehistoric animals rushing toward a cave for shelter. This is a reference to Cuvier's explanation of why the fossils of such monsters were so often found in caves as being because that's where they had been trapped at the time of the flood.

Martin was rewarded for his discipleship. He was applauded by Cuvier when the great man encountered the English artist's work in engraved form in Paris. When Cuvier subsequently came to London, he visited Martin's studio several times and complemented the artist on the scientific accuracy of his *Deluge*..

Cuvier's approval of Martin is further proof of the interrelationship between natural history and the visual arts at this time. Both were popular subjects. Both had a need for each other. The fact that Martin could illustrate and reinforce Cuvier by his stupendous images was a matter of consequence.

It is hardly surprising, therefore, to find Martin being cultivated by an English follower of Cuvier, Dr. Gideon Mantell. Mantell was a medical doctor and expert on comparative anatomy. He was able to piece together with particular brilliance reconstructions of prehistoric monsters. His masterpiece in this field was the discovery—or perhaps one should say the creation—of the Iguanodon.

As well as being an able paleontologist, Mantell was also a compulsive showman, living in high style and giving brilliant public lectures. His bestseller *The Wonders of Geology* (a record of his lectures) was deliberately designed to make clear "the wonders which modern geology has brought to light" and the ability of these to provide "more correct ideas of the mercy, wisdom, and goodness of the creator."[36] It probably did more to popularize geology and paleontology than any other publication of the time.

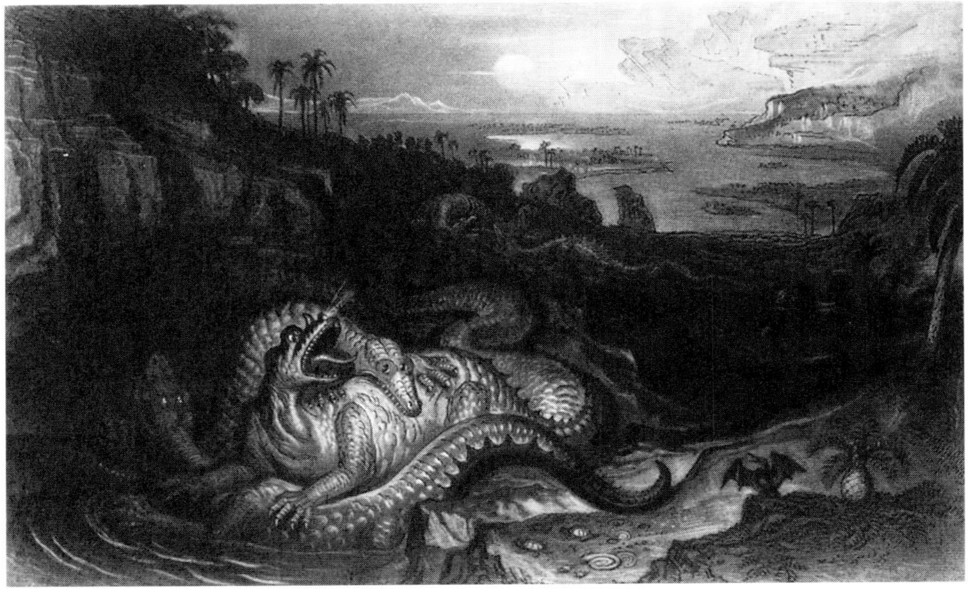

Figure 1.11. John Martin (England, 1789–1854), *The Country of the Iguanodon*, ca. 1837, engraving, frontispiece to Gideon Mantell, *Wonders of Geology*, vol. 1 (London, 1838).

Figure 1.12. *Dinner in the Iguanodon Model at the Crystal Palace on 31st December, 1853*, wood engraving from *Illustrated London News*, Jan. 7, 1854.

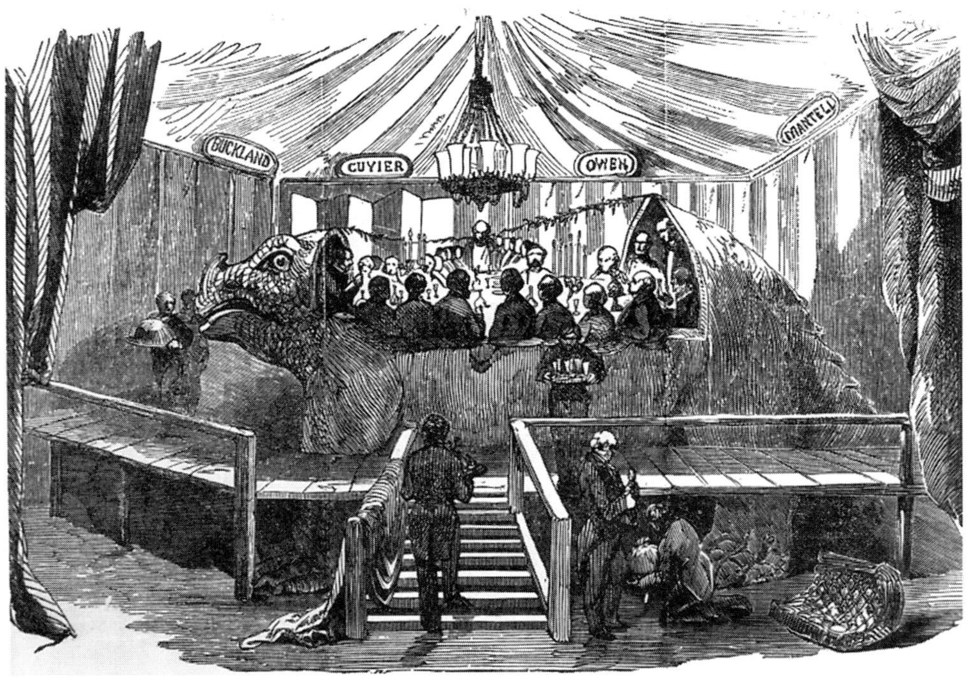

Mantell enlisted the services of Martin, the painter renowned for his Cuvierian visions of the history of the earth. When Martin came to visit his home to view the reconstruction of the Iguanodon, Mantell was touched and intrigued. "I wish I could induce him to portray the country of the Iguanodon," he wrote in his journal. "No other pencil but his should attempt such a subject."[37]

Martin was duly induced. The result was a marvelous conglomeration of fierce, dragon-like creatures. Mantell subsequently used it as a frontispiece to the *Wonders of Geology* (fig. 1.11).

Mantell was much pleased with the conception. We might be less so, simply because the creature Mantell named the Iguanodon is now thought to have been a herbivore. Its behavior in Martin's picture is hardly compatible with anything that might be conceived of as grazing. But of course it was necessary for conventionally Christian paleontologists in the early nineteenth century to portray prehistoric monsters as villainous. They were, after all, the animals that had failed to get into Noah's ark; there must have been something wrong with them. They were part of the evil of the prediluvian world that God had washed away with his flood. Sir Richard Owen—the first director of the Natural History Museum in London—stamped this characterization on them for good by naming the whole group "dinosaurs," a term deriving from Greek and meaning "terrible lizards." Years later the facade of the Natural History Museum bore out Owen's view; for it is arranged as a kind of "last judgment" with the lost prehistoric monsters on one side of the main entrance and the saved species still in existence

Figure 1.13. John Martin (England, 1789–1854), *The Great Day of His Wrath*, 1852, oil. The Tate Gallery, London

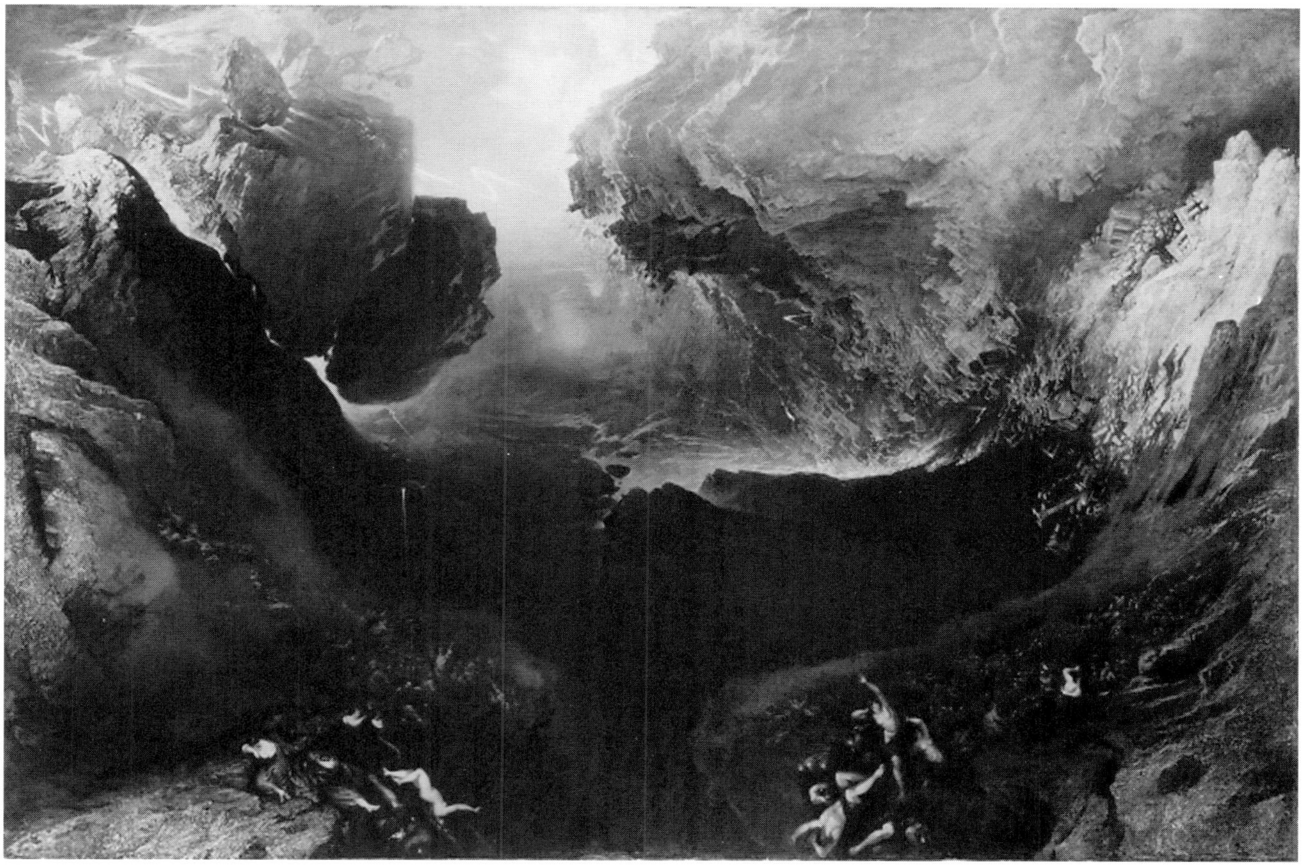

on the other.[38] Already in the 1850s, however, the popularity of the wicked monsters was such that they were appearing in places of recreation and entertainment. As part of the restructuring of the Crystal Palace after the Great Exhibition, B. Waterhouse Hawkins was asked to make models of the evil creatures from the past. He did so with the advice of Owen. When he completed his model of the Iguanodon in 1853, he and Owen and other distinguished professors of paleontology celebrated by having a meal inside it—rather like big game hunters boasting their bag (fig. 1.12).

Martin appears to have remained a Cuvierian to the end. The history of the earth was marked by catastrophes in which the wicked got their comeuppance. Right at the end of his life he impressively interpreted the Last Judgment in a trilogy of landscapes. One of these, the *Great Day of His Wrath* (fig. 1.13), illustrates the account of the end of the world to be found in the Book of Revelations. In Martin's picture it takes on the character of a Cuvierian natural disaster.

Ruskin and the Pre-Raphaelites

THE SUCCESS OF MARTIN'S PICTURE (the trilogy toured for thirty years after it was painted) might suggest that all was well with the alliance between art and geology. This might have been so at a popular level. But amongst scientists it was becoming increasingly hard to reconcile geological findings with the account of Creation in Genesis. From their point of view both Cuvier and Martin were hopelessly out of date. There is much evidence to suggest that most sophisticated observers found the whole issue to be of little significance or consequence. But there were many non-scientists and amateurs who found it hard to be so sanguine about the matter. For them the period of the 1850s was a time of

Figure 1.14. John Everett Millais (England, 1829–1896), *Ophelia*, 1852, oil. The Tate Gallery, London

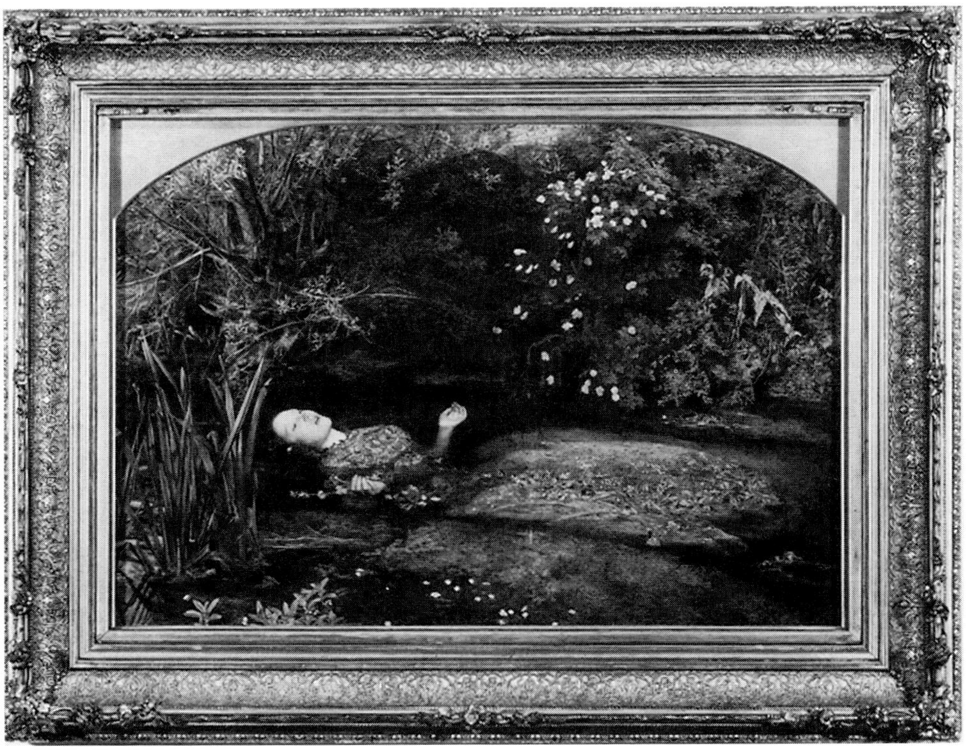

gathering crisis. As early as 1851 Ruskin was sensing the changing mood. In this year he wrote to his friend Sir Henry Acland, "If only the Geologists would leave me alone I could do very well. But those dreadful hammers!—I hear the chink of them at the end of every cadence of the Bible verses."[39]

Significantly, the 1850s was the period in which Ruskin experienced a crisis of faith. He resolved this crisis later in life partly by dismissing the conclusions drawn by professional geologists. During the 1850s he seemed to become obsessed with the idea of proving the artist to be a more revealing recorder and interpreter of the structure of the earth than the scientist. It is striking that the section of writing in which he praised Turner for his treatment of aiguilles (mentioned above) should have come from Volume Four of *Modern Painters*, which was published in 1856 at a time when his crisis of faith was reaching a climax.[40] The section on mountains is a riposte to contemporary geologists and an exhortation to artists to continue the work that Turner—now dead—had been doing.

Ruskin was at the same time looking for an artist to carry on this increasingly important mission. For a few years he thought he had found his man in John Everett Millais, the young Pre-Raphaelite painter whose technical skill and ability to record nature he found confirmed in the picture of *Ophelia* (fig. 1.14). The picture represented a remarkable blend of poetry and truth, showing as it did the tragic Shakespearian heroine being dragged down to her death in a real Surrey stream that the artist had painted on the spot. The mastery of observation and mood attracted Ruskin. But he thought Millais needed the experience of grander natural phenomena to develop his nascent vision. It was with this in mind that he took the artist with him to Glenfinlas in the Scottish highlands. Millais was set to paint a portrait of the critic as he stood before a particularly

Figure 1.15. John Brett (England, 1831–1902), *Glacier of Rosenlaui*, 1856, oil. The Tate Gallery, London

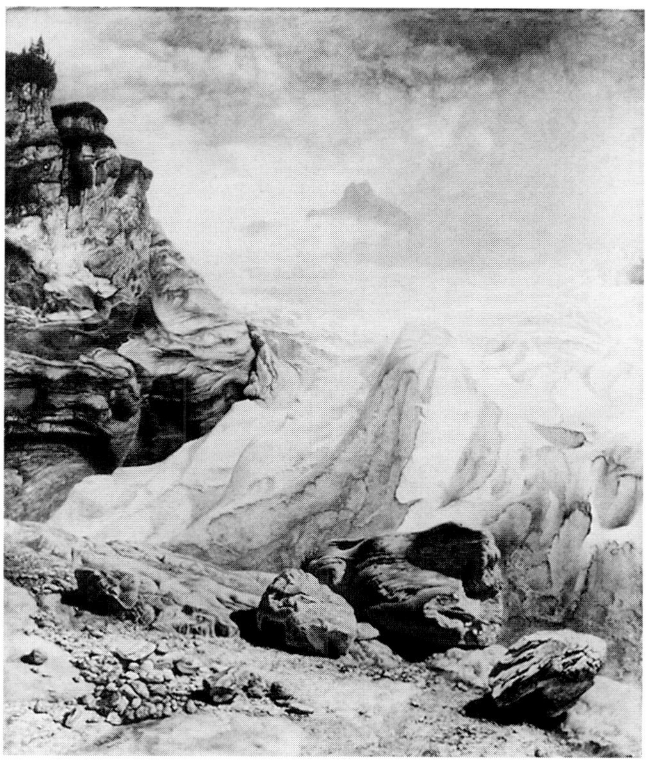

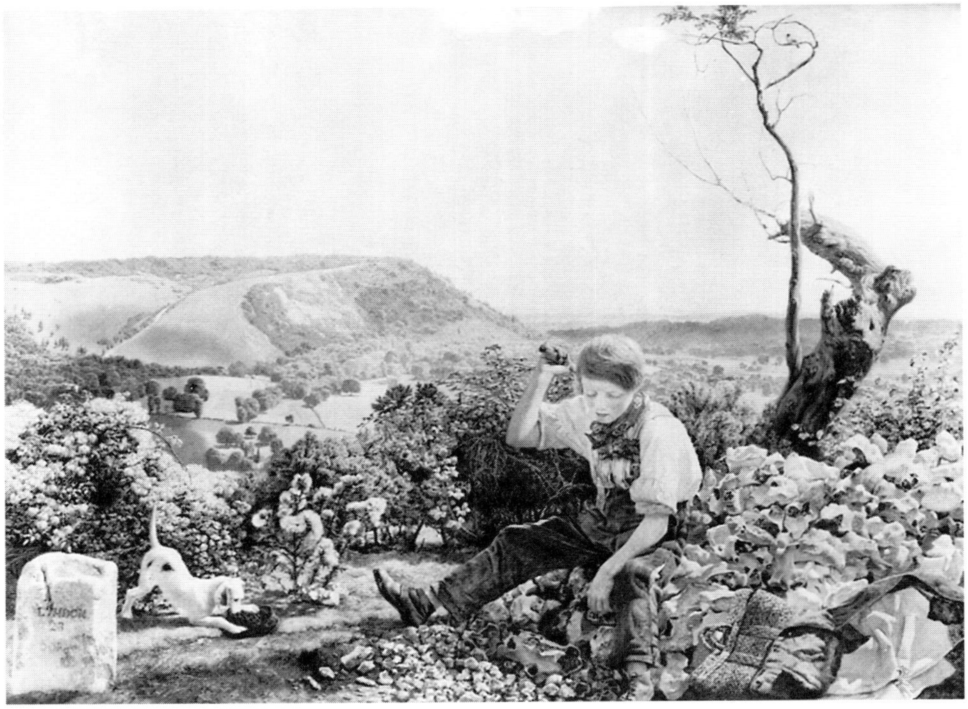

Figure 1.16. John Brett (England, 1831–1902), *The Stonebreaker*, 1858, oil. Walker Art Gallery, Liverpool

instructive formation of Gniess. The resultant image of the "prophet of the rocks" is hardly Turnerian in breadth or persuasive as a record of a moment of spiritual uplifting. Indeed—as is known to all students of Victorian art—the trip seems to have been inspirational to Millais in quite a different way. For it led to him developing an attachment to Ruskin's wife, the woman whom he was later to marry.

Despite this setback, Ruskin did not give up his search for a visionary geologist amongst the ranks of painters. Throughout the late 1850s he was scrutinizing the younger generation to find a man of the right caliber. His most promising candidate was John Brett, whose *Glacier of Rosenlaui* of 1856 (fig. 1.15) and *Stonebreaker* of 1858 (fig. 1.16) showed an unusually enlarged faculty for the description of rock formations. Brett seems, in fact, to have been inspired by Volumes Three and Four of *Modern Painters* to concentrate on such work.[41] But Ruskin felt that Brett, too, was failing to infuse sufficient inspirational grandeur into his work. Having visited the Alps in 1856, Brett had already been exposed to high mountains. But clearly he had not been in the right frame of mind. To make sure that he would see these inspirational forms in a different light, Ruskin took him in 1858 to the Val D'Aosta and gave him constant advice while he was painting the view.

Ruskin praised the resultant work in public, but seems in private to have felt some disappointment. For while Brett has depicted a somewhat grander spatial sweep than that in his two former "geological" landscapes, his technique remained restricted and meticulous. His choice of view (or rather of motifs—the picture is not a precise record of the view from a particular site) emphasizes the rural life of the valley, rather than communication with the heroic, uplifting spirit of the wild mountains.

Brett's "failure" is particularly interesting, because it did not come from an indifference to geology. Unlike Millais, Brett was keenly interested in the subject.

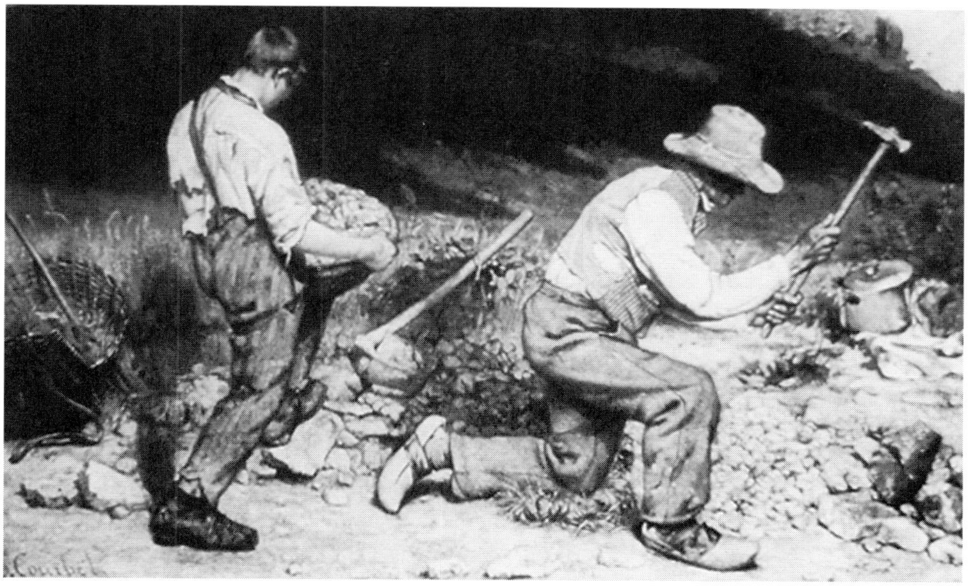

Figure 1.17. Gustave Courbet (France, 1819–1877), *The Stonebreakers*, 1849, oil, destroyed (formerly Gemaeldegalerie, Dresden).

His knowledge of it went far beyond what he could have read in Ruskin, and he went on painting landscapes with marked geological features throughout his life. The significant point is that Brett could not share Ruskin's concept of the study of rocks leading to a sense of the Divine. Brett was very much a part of a younger—and more informed—generation who simply did not choose to see such links. Indeed, as much as he did invest emotion in geological features, these tended to be of a negative kind. This is indeed discernable in the *Stonebreaker*. We should not be deceived by the prettiness of the boy or the brightness of the colors. The subject is a melancholy one. It was essayed by many artists of the period who wished to comment on the injustices of society. The most famous example was that by the French realist Gustave Courbet, the now destroyed *Stonebreaker* that was painted in 1849 (fig. 1.17).

Brett's picture shows a young boy whose life is already condemned to the soul-destroying manual labor of stonebreaking, an activity that was widely accepted at the time as one of the most degrading forms of physical labor currently being practiced. While the boy toils at his monotonous, unending task, his dog plays with his hat, alluding to an idea of the play and enjoyment that a child of his age should be allowed. In this context the geological forms evident in the picture take on a menacing aspect. We know from John Brett's own notes that he intended the scenery—with its barren chalk and hard, unyielding flints—to suggest "the wilderness of this world—outside Eden."[42] Geology here is no longer the solace of mankind, the material evidence of the presence of Divine benevolence. It suggests instead punishment and rejection and the forces of nature gradually wearing man down, making his toil seem fruitless. Small wonder that Brett, when confronting the Val D'Aosta, should prefer to distance the deceiving grandeur of the mountain and dwell instead on the thriving community below.

In the case of Brett, the artist's message has to be reconstructed by close reading and by the use of documentary evidence. Perhaps this is a sign of his limitations as a visual communicator. Yet he is a useful indicator of the contemporary predicaments. His position is a common one amongst those young landscapists who fell under the spell of Ruskin and the Pre-Raphaelites in the 1850s.

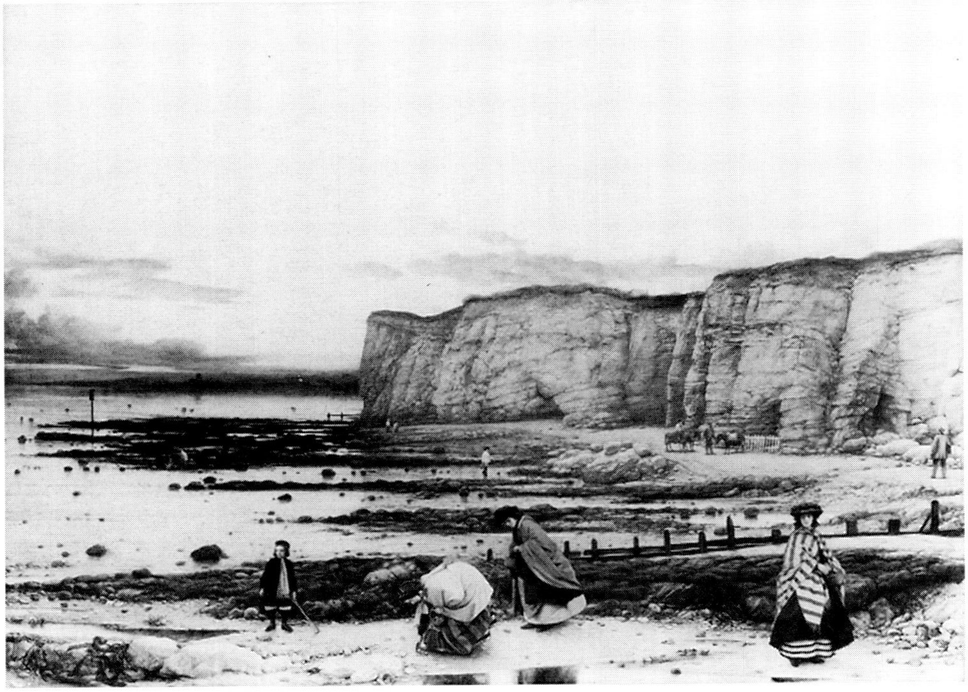

Figure 1.18. William Dyce (Scotland, 1806–1864), *Pegwell Bay*, 1860, oil. The Tate Gallery, London

Aesthetically, they fully shared an enthusiasm for the minute depiction of nature. Yet they could not read into that nature the ecstatic meaning that the Sage regarded as its ultimate significance. Frequently this lead to a kind of nullity. Nature is there, obsessively so, yet there is no sense of an emotive response to it.

Only one of these "Ruskinian" landscapes seems charged with strong feeling, and this feeling very much bears out the new, guarded approach to geology that Brett had sought to convey. That picture is *Pegwell Bay* (fig. 1.18 by the Scottish artist William Dyce. Dyce was then an older painter—born in 1806—and is known largely for his impressive but rather severe religious paintings, for example, the *Madonna and Child* painted for the Prince Consort in 1845 (fig. 1.19). A devout Anglo-Catholic, his principle mission in life was the revival of a "Christian" art that derived its inspiration from that of the late Middle Ages. In the 1850s, however, he had become impressed by the paintings of the Pre-Raphaelites. It was he, in fact, who first brought them to the attention of Ruskin.[43] Inspired by the Pre-Raphaelites, he began to add to his repertoire a series of meticulously observed depictions of British landscapes. *Pegwell Bay* is one of these. Unusually for Dyce, it has a pervasive poignancy. Unique in Dyce's oeuvre, it appears to have a special meaning. Superficially, it records a holiday he spent at the seaside; his family can be seen picking up shells in the foreground. Yet what is so remarkable is the dislocation between this frigid family holiday and the melancholy coast beyond. This is not the cheery seaside of the bourgeois holiday maker—the kind celebrated by William Powell Frith a few years earlier in his *Ramsgate Sands* (fig. 1.20). Dyce's picture shows a barren, friendless world. In her article on this picture in *Art History*, Marcia Pointon presents a brilliant reading of this work.[44] Her article suggests that the picture was essentially a disquisition on time. The quality of disquiet conveyed in it was a response to the new understanding of the scale of time resulting from recent scientific discoveries. Pointon suggests that the meticulously depicted chalk cliffs—

in which different strata can be seen clearly—are a direct allusion to geology. Dyce himself is shown facing them; he is the small figure in the middle ground on the extreme right of the canvas. Astronomy is also evoked in the painting. The full title of the picture alludes to the scene as a "recollection" of a precise time—October 5, 1858. This is the date at which Donati's comet could be seen most clearly in the south of England, and the comet is just visible in the center of the sky. Though it is hard to measure such things exactly, it seems that the small figure of Dyce is looking beyond the chalk cliffs to the comet. Taken together, cliffs and comet stand for Geology and Astronomy, which Tennyson later characterized as the two "terrible muses" of the age.[45] These were the sciences that seemed to have cut man off from the certainties of Genesis, making him feel powerless and isolated, surrounded by vast, immeasurable eons of time and space. The figure of Dyce in the painting sees this, and is himself almost faded to nothing while he has this vision. His frail, tiny form provides a poignant contrast to that image so prevalent in romantic landscape—that of the spectator staring in elation from the foreground at the enchanting scene. Meanwhile, as Dyce experiences this personal annulment, his family goes on picking up seashells in the foreground, oblivious of the new message that can now be read into them, the symbols of their precarious position on the edge of nothingness.

Figure 1.19. William Dyce (Scotland, 1806–1864), *Madonna and Child*, 1846, oil. Osborne House, Isle of Wight; Her Majesty the Queen

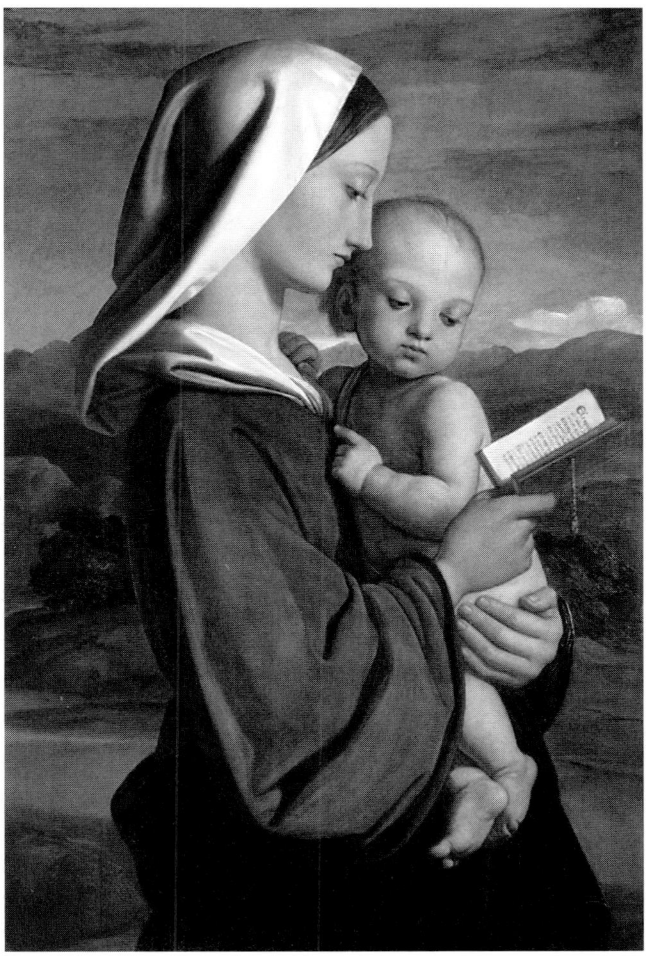

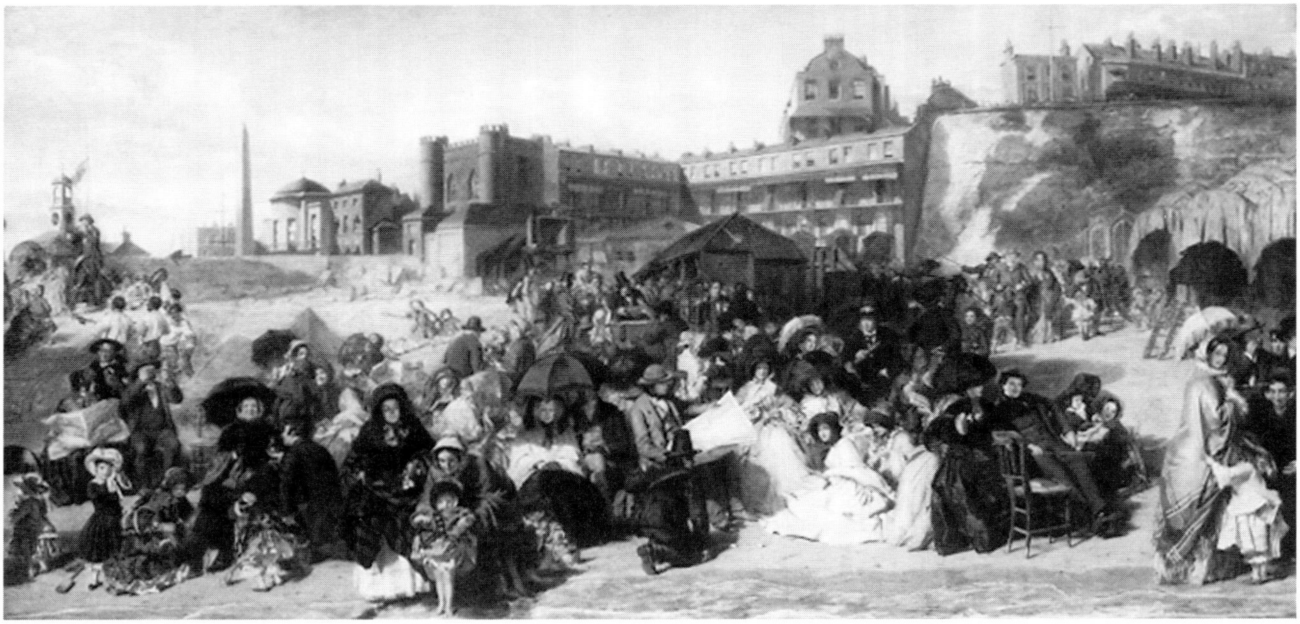

Figure 1.20. William Powell Frith (England, 1819–1909), *Ramsgate Sands*, 1854, oil. Her Majesty the Queen

This analysis—which I have perhaps dramatized a little more than can strictly be allowed from its presentation in Pointon's article—is highly persuasive, but it throws up an apparent paradox. This is that Dyce remained—from all the evidence—a deeply religious man. Dyce is feeling pain and anxiety at the loss of that Romantic enjoyment of the sense of the Divine in nature. On the other hand, just because nature and science can no longer reaffirm faith in a direct manner, does not mean that faith has ceased to exist.

It is possible, in fact, that this conflict of evidence may not have been as critical for Dyce as it was for some English artists. It was noted near the beginning of this chapter that there was a distinct difference of emphasis between the "Edinburgh" or "Scottish" school of geology and that of the "Oxford" or "English" school. It has been a particularly English obsession to seek to relate the findings of geology to religious faith, and a particularly English problem to see faith as being predicated upon evidence received from the world of nature. Dyce was, as Pointon has reminded us, a man of science as well as an Anglo-Catholic. He had won a prize for an essay on electromagnetism in 1830 from the University of Aberdeen, his alma mater.

Dyce's early training in science in Scotland may be significant here. For he may have been more interested in the "philosophical" issues that geology raised than in using it as a means of testing faith. It is striking that in his own comments on geology he talks about such mechanical features as the effects of erosion but says nothing about earth history or evolution.[46] Significantly, *Pegwell Bay* is a picture about time, but it is not a picture about history.

As Alan Staley has made clear in his book *The Pre-Raphaelite Landscape*,[47] the greater realism, and also something of the lyricism, to be found in Dyce's later landscapes owes a lot to Pre-Raphaelite landscape painting. But he kept clear of their literalism. However much he might have painted from detailed records he did not follow them in their obsessive *plein air*-ism. Nor did he show an interest in literal religious landscapes along the lines of those produced by William Holman Hunt, such as his *Scapegoat* (fig. 2.13). Indeed, in this sphere

Dyce moved rather in the opposite direction. For in the 1840s, when painting pictures such as *The Meeting of Jacob and Rachel* (fig. 1.21), he had been keener on giving a sense of the location of the Holy Land than he was now. In later life, Dyce seems to have felt that literal and religious truths were not coextensive in the way that Hunt—as will be seen in the next chapter—did. Hunt, it must be remembered, was a low churchman with Protestant inclinations; Dyce was a high churchman with Catholic ones. For Hunt physical and spiritual truth had to be coextensive. This was not the official Catholic line—as Cardinal Newman was to make clear later in his *Apologia* [48]—and it does not seem to have been Dyce's line either. A high church Anglican, he seems to have leaned in this instance towards a Catholic mode of thought. He was happy to have biblical events taking place in British landscapes, no doubt bringing them together in poetic union. Dyce was not a follower of Hunt in his attempted literalism. But he was a sufferer from it, because his later works were criticized from the point of view of the expectations that Hunt's biblical scenes had set up.

When Dyce's *Man of Sorrows* was exhibited in 1859, F. G. Stevens—former Pre-Raphaelite artist and now art critic—wrote in the *Athenaeum*, "The picture is completely realistic in all respects ... but why—with all this literalness—not be completely loyal, and paint Christ himself in the land where he really lived?"[49]

Figure 1.21. William Dyce (Scotland, 1806–1864), *The Meeting of Jacob and Rachel*, 1853, oil. Kunsthalle, Hamburg

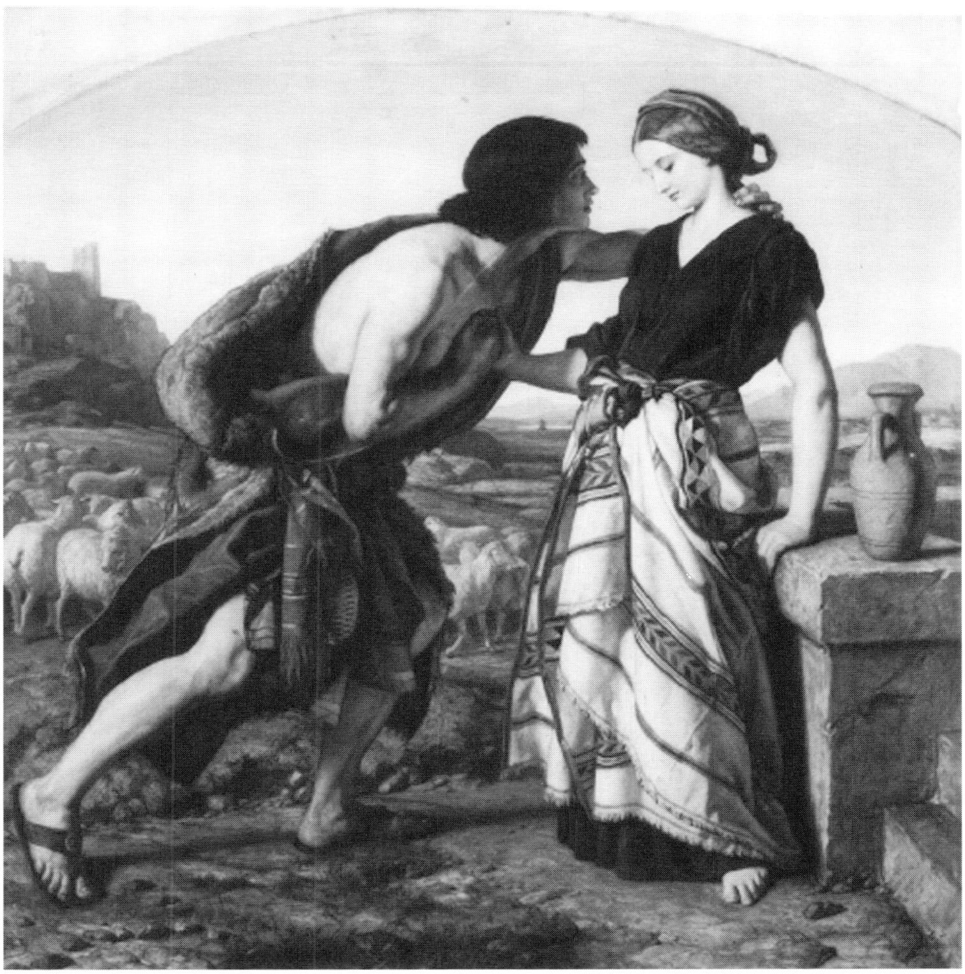

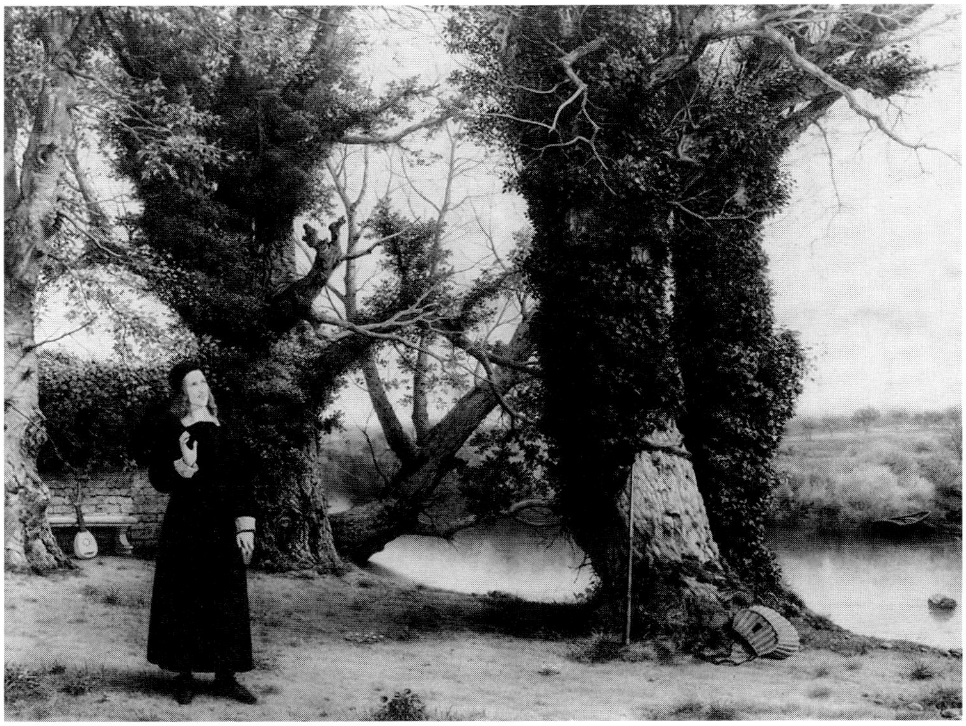

Figure 1.22. William Dyce (Scotland, 1806–1864), *George Herbert at Bemerton*, 1861, oil. Guildhall Library, City of London

It clearly might be objected that such literal contextualization was not necessary, since Dyce was wishing to show the isolation of Christ—something that comes out also in a quote from the Anglo-Catholic divine and poet John Keble that Dyce included in the catalogue.

> And when upon his drooping head
> His father's light was pour'd from Heaven,
> what time, unsheltered and unfed,
> Far in the wild his steps were driven,
> High thoughts were with Him, in that hour,
> Untold, unspeakable on earth.

It is the failure of earth to communicate, to be unable either to present or to receive the Divine message that obsessed Dyce. Dyce was keenly interested in rock formation and made precise geological notes on the different terrains that he visited—especially in his last years, when he made visits both to the Scottish Highlands and to Wales. But it was the "awful and terrific" aspects that obsessed him, as though these were symbols of the barrenness and degradation of the natural world.[50] While being an acute recorder of the natural world, he was also its critic. Possibly this explains the much commented upon anemia present in his late works. It is almost as if the earth is literally fading in significance. Certainly in the last years of Dyce's life he did a whole series of paintings of biblical scenes in which Christ is being shown in one kind of wilderness or meditative enclosure after another. There may well be a symbolic reason as well as a practical one for him showing these scenes in parts of the British Isles. This is that it emphasized that Christ's suffering was as relevant to people in the artist's own land and own century as it was to the people who had encountered Christ as a historical phenomenon.

Only one picture of this period appears to offer a more positive view—though appearances here, I will argue, are most definitely deceptive. The picture is *George Herbert at Bemerton* (fig. 1.22). It was exhibited at the Royal Academy in 1861, the last year in which Dyce showed any works.

The reputation of Herbert, one of the greatest devotional poets of the seventeenth century, underwent a well known revival in the mid-nineteenth century. Indeed, it was only in 1846 that a complete edition of his works was finally produced.[51] The revival of interest in Herbert received a great impetus from the ecclesiological movement, for which he represented a prototype. Dyce—a firm Anglo-Catholic—had visited Herbert's house, the parsonage at Bemerton, just outside Salisbury. Dyce was in fact a friend of Cyril Page, who was then rector of Bemerton and a devotee of his predecessor.

At first sight it might seem that Dyce's painting, with the image of Herbert musing in his garden, demonstrates that he had turned his back on the vision of the *Man of Sorrows* and *Pegwell Bay*. For in his portrayal of Herbert all is calm and quiet. Technically, the artist has returned to his meticulous, admirable, but not quite inspiring mode. Once again, there is a text to help us read the picture's theme. As with the *Man of Sorrows* this text directs our attention to a relationship between figure and landscape. As with the *Man of Sorrows*, too, the relationship is not altogether a happy one.

The text for this picture—the one that was printed in the Royal Academy catalogue when the work was exhibited—is a poem by Herbert. It is the one titled *Virtue* that is cited by Isaac Walton in *The Compleat Angler*.[52] The first verse opens with an idyllic evocation of a fresh spring day:

> Sweet day, so cool, so calm, so bright
> The bridall of the earth and sky

But there is a sting in the tail, for the second half of the stanza emphasizes the day's transience

> The dew shall weep thy fall tonight;
> for thou must die.

At the end of the poem Herbert contrasts this transience of the natural world with the eternity offered to the virtuous soul:

> Only a sweet and vertuous soul,
> like season'd timber never gives;
> but though the whole world turn to coal
> then chiefly lives.

Dyce is using Herbert's poem as a reminder that the natural world is not the appropriate place to lodge one's faith. It is the inner world that nurtures this, one that is beyond the reach of objective observation. This argument is striking because it leads away from the naturalistic debate of the time. It is, moreover, a view traditionally related to Catholic, rather than Protestant thought, and something that—as has already been said—one can find contemporaries like Cardinal Newman also stressing.

There are many features within the picture that reinforce the reading suggested by Herbert's poem. The lute that lays against the tree is an image of lyricism, the fishing basket a symbol of Herbert's pastoral role as a "fisher of men."[53] At the same time there are visual indicators of a rejection of the Romantic, pantheistic reading of nature. Herbert's position—turned away from the landscape—is the direct opposite of the motif common in Romantic landscape of a figure turning towards nature and deriving spiritual sustenance from the experience. If Herbert is deriving inspiration from anywhere, it is from the book he has been reading and in which he still has a finger. It might even be that Dyce is trying to emulate something of the "meditative" attitude to nature found in Herbert and in other metaphysical poets. In contrast to Romantic ecstacy, this mode of thought looks to nature as a vehicle for symbolic references to eternity that have to be carefully studied before their meaning becomes apparent. As we have seen, Dyce's picture is similarly "coded," leading through exploration to a meaning that is the reverse of what was at first apparently intended. There are possibly even intended affinities here. For, as has been remarked elsewhere,[54] Dyce's picture has something of the delicacy and emblematic character of an Elizabethan or early seventeenth century miniature. This kind of miniature is, of course, the nearest visual equivalent to the sonnets and symbolic short poems of the period. Iconically such miniatures might have set off Dyce's train of thought, particularly those that show a melancholic reflecting on nature, such as Isaac Oliver's portrait of Herbert's kinsman, *Lord Herbert of Cherbury* (fig. 1.23).

Perhaps this distinction between "romantic" and "metaphysical" responses to nature can be drawn out further by one final comparison. In Herbert's painting there is an apparent ironic play on the image of the church in the landscape as a symbol of religious security. Salisbury Cathedral can be glimpsed in the background, it is shown as pale and tiny—so tiny that it might almost be missed. This appears to be the inverse of the towering sublime image of Salisbury favored by the landscape painters of the early nineteenth century—such as Turner and Constable.[55] Indeed, there may even have been a deliberate play on Constable's *Salisbury Cathedral from the Bishop's Grounds* (fig. 1.24), a work that was well known at the time. Constable had shown the great cathedral viewed through a "gothic" arch of trees—a device that emphasized the harmonious relation of the religious edifice to the works of God in nature. The figures in foreground—turned in towards the scene in characteristic Romantic fashion—appear to appreciate this relationship. The fact that this view is the one obtained,

Figure 1.23. Isaac Oliver (England, 1568–1617), *Lord Herbert of Cherbury*, ca. 1600, oil. The Earl of Powis

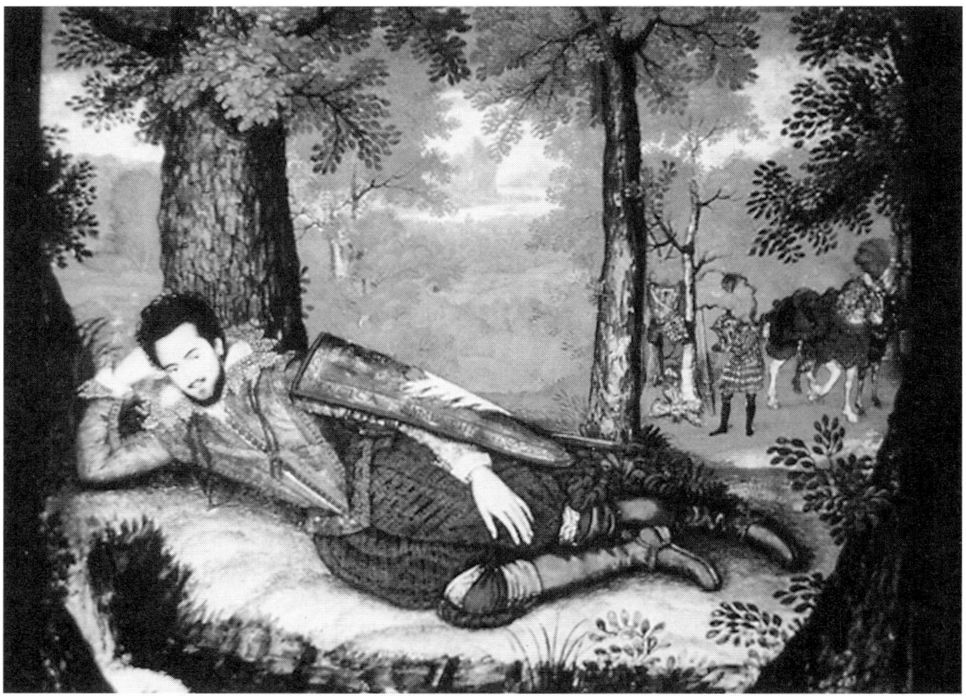

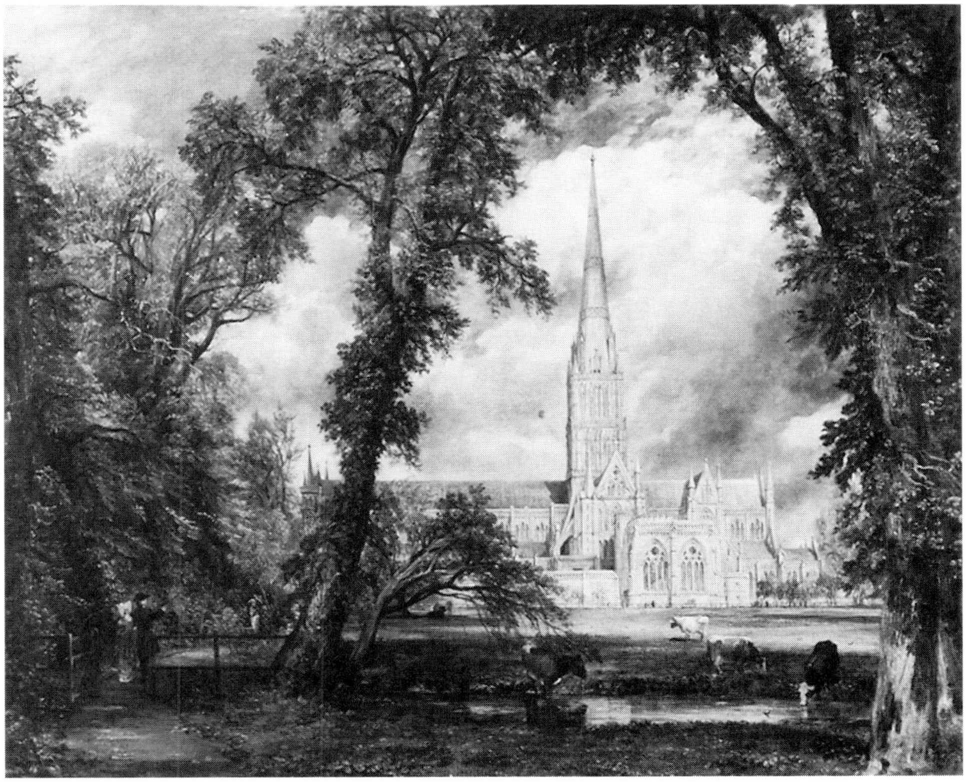

Figure 1.24. John Constable (England, 1776–1837), *Salisbury Cathedral from the Bishop's Grounds*, 1823, oil. Victoria and Albert Museum, London

as the title of the picture says, from the "Bishop's Grounds" underpins this vision as one natural to the Anglican order of religion—that religion which in 1824, the time when Constable painted the picture, was still complacent about its position, both socially and spiritually.

In contrast to Constable's "episcopal" view of Salisbury, Dyce has given that of the country parson. Further removed from the symbol of authority, he draws his faith from an inner world. As an Anglo-Catholic, Dyce did not take the "social" role of the church for granted, but was concerned with the raising of the spiritual temperature. For him the complacent "natural" image of Anglicanism no longer pertained. The fact that the reliance on the contemplation of nature as a means of reinforcing faith was being undermined at this time appears to have been for Dyce a welcome clarification of issues.

Toward Aestheticism

In this chapter I have been concentrating on the loosening of ties between naturalism and religious faith that was precipitated by scientific developments. I have also suggested that this created a "crisis" for artists like Dyce, Brett, and the Pre-Raphaelites who were trained in the tradition that naturalism was the principal goal of the artist who depicted nature. The responses they adopted provided a critique on naturalism without abandoning its language. Such a position, while often leading to curious and interesting results, was hardly conducive to the establishment of a new pictorial mode. It was a later generation—that of

Whistler and the Impressionists—who was to take that decisive step. This was the generation that Ruskin accused of having a lack of naturalistic concern, a fault that, as he made clear in his Oxford Inaugural Lecture,[56] he blamed on Darwin. These were the "aesthetes" that he railed against in later life.

In detail Ruskin may have been wrong, but in essence he was correct. For the new vistas opened up by science seemed to create less and less possibility for celebrating the wonders of nature in religious terms. Time and again one finds references amongst the "aesthetes" to the meaningless world that modern science has left us with. Indeed, the seminal essay of this generation, the conclusion to Walter Pater's *The Renaissance* of 1872, opens up with a frank acknowledgement of the changefulness that the modern world had created for itself: "To regard all things and principles of things as inconstant modes or fashions has more and more become the tendency of modern thought."[57] And he goes on to advocate a turning away from inquiry, a celebration of the momentary: "While all melts under our feet, we may well grasp at any exquisite passion, or any contribution to knowledge that seems by a lifted horizon to set the spirit free for a moment."

It was the celebration of such momentary effect that Whistler, the painter most closely associated with the aesthetic movement, specialized in. In his image of a falling rocket above the Cremorne Pleasure Gardens he caught the quintessence of the exquisite momentary effect. There is no need for him to scrutinize the rocks and cosmos for his insights. Nature, he once declared, is like a piano. The notes are there, but it is up to the artist to devise the chords to be played with them.

It was this cavalier attitude that caused Ruskin such pain and which goaded him into his celebrated attack on Whistler as a "Cockney...flinging a pot of paint in the face of the public."[58] For Ruskin nature was not a piano but a Bible, and one that scientists and the younger generation of artists were failing to read properly. The new aestheticism, the new science, were but different fragments of the same soullessness, a falling away of art and scholarship into fragmentary disciplines of episodic purposelessness.

We can take a more positive view of the matter, seeing how the drive towards greater specialization has produced immense result in both the arts and the sciences. We can also see how science continues to remain an inspiration to the visual artist, even when the bonds that linked them both to religious faith were severed. Yet we also know that the rewards of specialization have been won at a price. For the modern painter, as for the modern research scientist, there is a problem of communication; a problem of communication not just between himself and the public, but also, frequently, between himself and his colleagues. While in no way sharing their premises, we can perhaps look back with a little envy at the cultural world of the early nineteenth century, when art, religion, and science could be seen as part of some great continuum, sharing a unity of purpose that gave them a common, communicating voice.

Notes

1. See N. Rupke, *The Great Chain of History: William Buckland and the English School of Geology, 814–1849* (Oxford: Clarendon Press, 1983), 9, 18, 234, 271.

2. Lynn Barber, *The Heyday of Natural History, 1820–70* (London: J. Cape 1980). See especially p. 14.

3. See especially M. Pointon, "Geology and Landscape Painting in Nineteenth Century England," *Images of the Earth: Essays in the History of the Environmental Sciences*, ed. L. J. Jordanova and R. Porter (London: British Society for the History of Science, 1979), 84–118.

4. John Ruskin, *Modern Painters*, in *The Works of John Ruskin*, ed. by E.T. Cook and Alexander Wedderbum, 39 vol. (London: George Allen, 1905–1912) 5:377–78.

5. John Constable, lecture at the Royal Institution, 1836. See C. R. Leslie, *Memoirs of the Life of John Constable Esq., R.A.* (London: Phaidon Press, 1843), 333.

6. D. E. Allen, *The Naturalist in Britain* (Harmondsworth: Penguin, 1976); R. Porter, *The Making of Geology: Earth Science in Britain 1660–1815* (Cambridge: Cambridge University Press, 1977); Rupke, *Great Chain of History.*

7. Allen, *Naturalist in Britain*, 53.

8. Barber, *Heyday of Natural History*, 17–18.

9. Rupke, *Great Chain of History*, 5.

10. See T. Mitchell, "Caspar David Friedrich's *Der Watzmann*: German Romantic Landscape Painting and Historical Geology," *Art Bulletin* 66 (1984): 452–64. Geology and German landscape painting also is discussed in M. Cheetham, "Revision and Exploration: German Landscape Depiction and Theory in the late 18th Century" (Ph.D., University of London, 1982), 282–347.

11. See B. Novak, *Nature and Culture: American Landscape Painting 1825–75* (New York: Oxford University Press, 1980), especially 47–77.

12. Constable to John Dunthorne, 29 May 1802, *John Constable's Correspondence*, ed. R. B. Beckett, (Ipswich: Suffolk Record Society, 1964), 2:32.

13. See Leslie, *Life of Constable*, 298.

14. T.J. Chandler, *The Climate of London* (London: Hutchinson, 1965), 25; K. Badt, *John Constable's Clouds* (London: Routledge & K. Paul, 1950).

15. Basil Taylor, *Constable: Paintings, Drawings and Watercolours* (London: Phaidon, 1975), 28.

16. L. Parris and C. Shields, *John Constable* (London: Tate Gallery, 1985), 7.

17. Leslie, *Life of Constable*, 101.

18. Allen, *Naturalist in Britain*, 40.

19. Barber, *Heyday of Natural History*, 51.

20. Constable to John Fisher, 1 April 1821, *Constable's Correspondence*, 4:65.

21. Charles Lyell, *Principles of Geology*, 3 vols. (New York: Appleton, 1857), 1:127.

22. Constable's library is listed in Parris et al., eds., *John Constable: Further Documents and Correspondence* (London: Tate Gallery, 1975), 44.

23. Porter, *Making of Geology*, 142–43.

24. See for example the works of John Gage, *Colour in Turner* (London: Studio Vista, 1969).

25. See F. F. Cunningham, *Revolution in Landscape Science* (Vancouver, Canada: Tantalus Research, 1977), 74. I am most grateful to Justine Hopkins for drawing my attention to this information.

26. Rupke, *Great Chain of History*, 40–41.

27. Ruskin, *Praeterita*, in *The Works of John Ruskin*, 35:115.

28. Ruskin, *Modern Painters*, in *The Works of John Ruskin*, 6:237.

29. Ruskin, *Modern Painters*, in *The Works of John Ruskin*, 5:377–78.

30. This reading is made clear by the lines from Turner's poem "Fallacies of Hope" that were appended to the exhibition catalogue entry for the painting. See M. Butlin and E. Joll, *The Paintings of J. M. W. Turner* (New Haven, Conn.: Yale University Press, 1977), 79.

31. For this and other general biographical details of Martin's life I have relied on: W. Feaver, *The Art of John Martin* (Oxford: Clarendon Press, 1975); and M. L. Pendered, *John Martin. Painter. His Life and Times* (London: Hurst & Blackett, 1923).

32. Feaver, *Art of John Martin*, pl. 4.

33. The knowledge of Cuvierian geology in England was based on Robert Kerr's translation of this work (1813). See Rupke, *Great Chain of History*, 39.

34. Pendered, *John Martin*, 133.

35. This feature is also shown, incidentally, in the *Deluge* painted by Francis Danby in 1840, now in the Tate Gallery. I am most grateful to Justine Hopkins for drawing my attention to this fact.

36. G. Mantell, *The Wonders of Geology*, (London: Relfe and Fletcher, 1838), 7.

37. E. Cecil Curwen, ed., *The Journal of Gideon Mantell* (London: Oxford University Press, 1940), 125; cited in Feaver, 146.

38. See Mark Girouard, *Alfred Waterhouse and the Natural History Museum* (London: Yale University Press in association with the British Museum (Natural History), 1981), 57.

39. Martin to Sir Henry Acland, 24 May 1851, in *The Works of John Ruskin*, 36:115.

40. See p. 10, this manuscript.

41. See entry by Chris Mullins for *Glacier at Rosenlaui*, *Zwei Jahrhunderte Englische Malerei* (Munich, Germany: Haus der Kunst, 1979), 376.

42. See K. Kajander, "John Ruskin's Pre-Raphaelites; John Brett and John William Inchbold" (unpublished M.A. report, University College London, 1975), 31, pl. 12. Kajander's report draws upon notes and sketches in the collection of Brett's descendants.

43. Ruskin mentioned this in a letter of 28 December 1882 to Ernest Chesneau; see John Ruskin,"Letters & Correspondence," *The Works of John Ruskin*, 37:427–28.

44. Marcia Pointon, "The Representation of Time in Painting: A Study of William Dyce's *Pegwell Bay*; A Recollection of October 8th, 1858," *Art History* 2 (1979): 99–103.

45. In the poem "Parnassus," first published in 1889 (Globe Edition *Poetical Works of Alfred Tennyson*, Edinburgh: Hay & Mitchell, 1899, 609).

46. See for example the passage in which Dyce compares the effects of erosion on Scottish granite in contrast to that on Welsh slate; quoted in A. Staley, *The Pre-Raphaelite Landscape* (Oxford: Clarendon Press, 1973), 166.

47. Ibid., 162.

48. John Henry Newman, *Apologia Pro Vita Sua* (London: Sheed and Ward, 1973). See especially pt. 7, "General Answer to Mr. Kingsley" (Doubleday Anchor Edition, 1956), 334 ff.

49. *Athenaeum*, 11 May 1860, 653. Quoted by Staley, *Pre-Raphaelite Landscape*, 165.

50. Staley, *Pre-Raphaelite Landscape*, 163.

51. *The Works of George Herbert in Prose and Verse*, 2 vols. (London: William Pickering, 1846) 2:324. The Pickering edition contains an engraving of the portrait of Herbert by R. White (1674) that was used by Dyce as the basis for his representation of the Divine. White's portrait, however, was reproduced in many nineteenth-century editions of Herbert's works.

52. Izaac Walton, *The Compleat Angler*, Everyman Edition (Oxford: Clarendon Press; New York: Oxford University Press, 1983), 97.

53. This aspect of Herbert is emphasized by Isaac Walton. Sometime in the nineteenth century myth arose that the basket originally had been included by Dyce because at first Walton had been shown fishing in the picture, and that Dyce had later painted him out when he had discovered that such a reference would be an anachronism. This story, which has been retold by most twentieth-century commentators on the picture (see Staley, 169), is clearly an absurdity, for several reasons. First, the inclusion would have been no anachronism, since Walton was of the same generation as Herbert and had known him personally (even though *The Compleat Angler* was published after Herbert's death). Second, there are no visible signs of overpainting in the picture. Third, such an inclusion would have made compositional and thematic nonsense of the work. Finally, there is nothing inappropriate in the depiction of fishing tackle in a picture of Herbert. For as well as being a "fisher of men" the poet was, according to Walton, also a keen angler.

54. Staley, *Pre-Raphaelite Landscape*, 169.

55. See, for example, Turner's watercolor of *Salisbury Cathedral from the Cloisters* (1802) in the Victoria and Albert Museum, London.

56. Ruskin delivered the Oxford Inaugural Lecture in 1870. See Ruskin, *Lectures on Art*, in *The Works of John Ruskin*, 20:17.

57. Walter Pater, *The Renaissance* (World Library Edition; Oxford: Oxford University Press, 1986), 150.

58. Ruskin, *Fors Clavigera*, in *The Works of John Ruskin*, 29:159.

II. Religious Art and "Agreeable Reality"

"It is the difficulty of giving agreeable reality to sacred subjects which daunts the modern artists, living in a critical age and sensitive to criticism."

—*J. E. Millais*[1]

THE DOUBTS CONCERNING the nature of spirituality that beset the mid-Victorian world had repercussions—not surprisingly—for the devotional art of the period. Dr. Lindsay Errington's study[2] has made us aware of the extent to which contemporary religious controversy affected such work. I do not wish to resurvey that well-covered ground here. Rather, I want to look in more detail at a specific and largely pictorial issue: the way in which artists set out to align rationalist inquiry into the history of the Christian religion with the depiction of sacred subjects.

In the case of religious paintings the scientific investigations that threatened most directly were those of the new biblical scholars who threw doubts upon the truthfulness of the most hallowed scenes of Christian art. Their investigations cast new light on the age-old problem of the degree to which historical accuracy should be pursued in religious imagery. To a large extent such onslaughts came to British artists from abroad—in particular France and Germany—and had already been the subject of debate amongst artists in those countries. Certain French and German pictures provided important reference points for painters in Britain.

It is in the works of the Pre-Raphaelite painter William Holman Hunt that these problems manifested themselves with the greatest urgency. Although I intend to look at certain artists that preceded him—in particular the Scottish painter William Dyce—I would like to begin by considering the contemporary reactions to Hunt's *The Light of the World* (fig. 2.1), perhaps the most compelling of all Victorian religious pictures. Exhibited at the Royal Academy in 1854, it shows Christ knocking at the door of the human soul. The subject was inspired by a quotation from the Book of Revelations (3:20): "Behold I stand at the door and knock; if any man hear my voice and open the door, I will come into him, and will sup with him, and he with me."

The picture is, of course, an allegory. Yet what is so curious—and what helps to make the picture so arresting—is the degree of realism that Hunt has brought to its execution. In recent years it has been argued that such "realistic" symbolism relates to current typological practice in religious thinking, particularly amongst the low church evangelical group, for whom Hunt had great sympathy. In his book *William Holman Hunt and Typological Symbolism*, George Landow demonstrates how Hunt was particularly attracted to this mode of thought as it appears in the writing of John Ruskin.[3] The fundamental concept of typology, whereby one "real" event prefigures another, certainly accorded with the more general tradition of using realistic details to point symbolically towards ideas and characters that Hunt absorbed from the tradition of Hogarth. Furthermore, it was important in showing that such realistic treatment could have a spiritual dimension. It was Ruskin's emphasis on this fact that gave a particular authority to his call for "truth to nature" and to its interpretation by the Pre-Raphaelites as meaning the minute observation and elaboration of detail. Such detailed "truth" was a characteristic of *The Light of the World*. In his autobiography and history of the Pre-Raphaelite movement, Hunt goes to great lengths to explain the trouble he took to depict the light effects and objects in the picture directly from nature. Much of the background was painted in an orchard between eight p.m. and four a.m. at times of full moon. Nothing appears to have

Figure 2.1. William Holman Hunt (England, 1827–1910), *The Light of the World*, 1853, oil. Keble College, Oxford

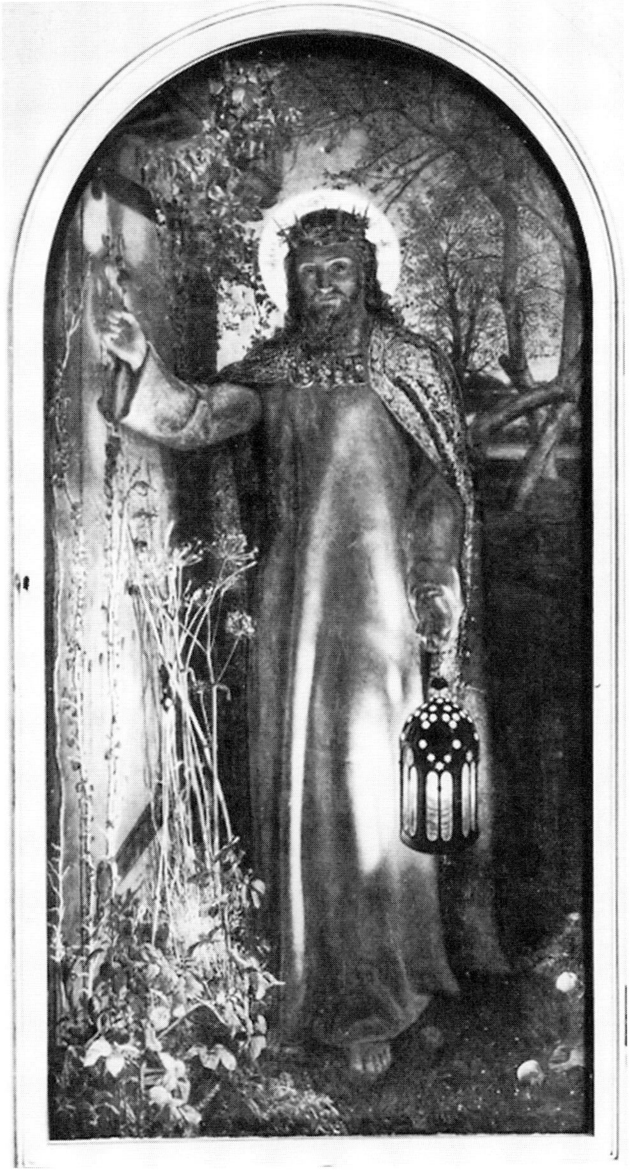

Figure 2.2. After Philipp Veit, *Christ Knocking on the Door of the Human Soul*, 1824, engraving.

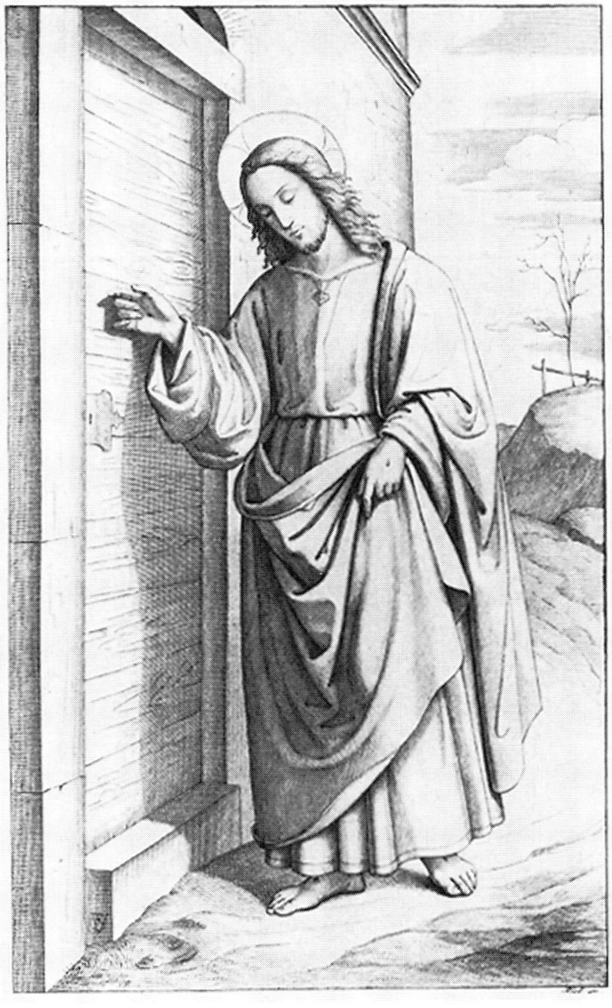

been left to fancy. He even had the lamp Christ is holding specially constructed according to his own design so that it could be painted directly from the real object.

In this work, in fact, Hunt is trying to combine realism and traditional features of religious art. And, as so often happens in the case of compromise, he fell foul of the supporters of both activities. For traditionalists, he appeared to have committed unheard of heresies. This point of view was represented at the time by the distinguished scholar Dr. Gustave Waagen, director of the Berlin Museum. In a letter to *The Times* Waagen complained that Hunt had confused the iconography of two separate roles of Christ, that of Christ the King (represented by the rich overmantle and golden crown) and that of Christ the Martyr (evidenced in the crown of thorns).[4] Waagen's objections, which stemmed from his knowledge of medieval and Renaissance art, were made partly because of his familiarity with the works of the principal German religious revivalist painters of the time, the Nazarenes. The difference between Hunt's approach and that of the German artists can be gauged when one looks at the Nazarene engraving that Hunt was accused of plagiarizing in his picture: Philipp Veit's *Christ Knocking on the Door of the Soul* (fig. 2.2), in which there is no iconographical confusion. Furthermore, the comparison of these two works helps to demonstrate Waagen's second objection to Hunt's painting. This was the conflict that existed in it between symbolical and realistic representation. Waagen pointed out that it was irrational to include a halo, which is a schematized symbol for light, in a picture that set out to treat light naturalistically.

Hunt found no difficulties in brushing such objections aside. He saw no reason why he should refrain from using traditional iconography in a novel way, and felt realism of treatment was essential to bring the message of the work home to a modern audience. Like the evangelical preachers that he so admired—and who were so popular at the time—he wanted to emphasize the modernity and personal relevance of the Christian message. Even the halo could be defended—at a pinch—on the grounds that it had the color and appearance of a full moon, and could therefore be related to a natural phenomenon. Hunt, it should be remembered, had taken care to paint the background out-of-doors at the time of the full moon. Christ's halo is, in effect, the full moon magnified, a piece of nature revealing its fuller purpose as an emblem of the Divine.

However, if Hunt had no problems in countering the pedantic objections of the traditionalists, he was far more susceptible to the attacks of those who challenged his realism. Such attacks came from those who sought a literal truth in their pictures—religious or otherwise. The principal criticism from this quarter, which Hunt transcribed in his autobiography, was provided by the Scottish sage and historian Thomas Carlyle. When he saw *The Light of the World* in Hunt's studio, Carlyle dismissed it as a "papistic fantasy," remarking of the image of Christ, "If you realized his character at all you wouldn't try to make people go back and worship the image that priests have invented of Him to keep men's silly souls in meshes of slavery and darkness. Don't you see that you're helping to make people believe what you know to be false, what you don't believe yourself?"[5]

Carlyle's objections were one of the reasons why Hunt decided to pursue a radical new direction in his religious art, going out to Palestine to paint biblical subjects in their actual historical location. This may not have been quite what Carlyle had in mind when he exhorted Hunt to paint the "real" Christ. But Carlyle must surely already have been aware by this time of the attempts that were being made to reconstruct a historical knowledge of Christ that did not rely simply on a reading of traditional scriptural sources. Eight years before Hunt painted his picture in 1846, a German work had been published in England that had caused a furor by making just such an attempt. This was George Eliot's translation of D. F. Strauss's *Life of Jesus*, which had appeared in Germany in 1835 and had already run into four editions in that country by 1840.[6] In his book Strauss set out to separate what was mythical from what was historically verifiable in the accounts of the life of Jesus. Although Strauss's conclusions were later to be superceded, his attempt was important for initiating a new approach to biblical studies. A later commentator remarked, "Strauss' criticism broke down the ramparts of dogmatism, new and old, and opened to the inquiring mind the breach through which the conquest of historical truth might be won."[7] Even those who distrusted Strauss's conclusions (Carlyle was vituperative in his attack on the book) felt a new onus created by the work. It was beholden to them to consider more closely the historical facts of Christ's life in order to be in a position to reject Strauss's heretical assertions.

Hunt's championship of a realistic mode of painting made the question of the truthful representation of religious subjects a particularly important one to him. It is hard to say whether he knew of Strauss's book by the 1850s (he certainly did later). But he was definitely aware of a controversy that had been in existence since the 1830s about the degree to which artists should be concerned with historical and topographical accuracy in the portrayal of biblical subjects.

This controversy led back to artists working in Rome. On the side of tradition there had been the followers of the Nazarene artist Friedrich Overbeck. These artists firmly had rejected the notion that historical reconstruction had a place in religious art. Such art, they felt, should be based both iconographically and stylistically upon the traditions of the past. As a devout Catholic, Overbeck gave as much authority to the practices of the Church as to independent readings of the Bible. He gave special attention to religious art of the Middle Ages since he felt that it had derived its inspiration from the Church and thereby embodied revelations that could not be acquired in any other way. For the British this point of view was presented by Nicholas Wiseman, the priest who was later to spark off the controversy over the Catholic reconversion of the British Isles in 1850. In his *Four Lectures on the Office and Ceremonies of Holy Week* (1839) Wiseman wrote with great enthusiasm about the devotional purity and orthodoxy of Overbeck's art. He also used one of the German artist's pictures as a frontispiece to his work. He further advanced the opinion that the art from which Overbeck drew inspiration was modeled in its design and costume on the ceremonies of the Church. He cited as a particular example the frescoes by Fra Angelico in the St. Stephen Chapel in the Vatican.[8]

There were many—even amongst those who did not have Catholic sympathies—who were prepared to concede that the Nazarenes had adopted the right tone in religious art. Already in 1820 the English painter Charles Eastlake had written in a review of the Casa Bartholdi frescoes by Overbeck and his associates, whose subject was the biblical story of Joseph, "For simplicity, holiness and purity, qualities which are the characteristics of the scriptural scenes, no style was better adapted than that of the Germans. This style has little or nothing to do with reality; it diffuses a sort of calm and sacred dream."[9] To achieve truly spiritual works, it appeared necessary to abandon the materialism of the present and go back to the ethos of the "age of faith."

However logical this argument was, it presented severe difficulties for those artists—including Eastlake—whose manner of painting was of a more sensuous bent. Eastlake went through considerable soul searching on this point, and ended up by expunging devotional subjects for a time from his repertoire. He did not return to religious themes until the 1840s, when critical arguments on the subject were taking a new turn.

By this time an alternative to the Nazarene style was emerging. It stemmed from the unlikely source of Horace Vernet, the highly materialistic painter of journalistic accounts of Napoleonic battle scenes and other modern life dramas. Vernet was director of the French Academy in Rome from 1828 to 1835 and, perhaps because of his location, turned his attention to the problem of painting scriptural scenes at that time. His *Judith and Holofernes* (fig. 2.3), which was exhibited in 1831, astounded the Roman art world because it seemed to fly in the face of what had been accepted as the norm for biblical scenes. Instead of a quattrocento grace and medievalizing costume, there is a thoroughly modern treatment of detail and an attempt to reconstruct the period and location of the event. This, of course, was no more than what was being attempted for secular history pieces by such artists as Paul Delaroche in Paris, but transferred to a biblical subject such literalness seemed lacking in decorum. During the 1830s there persisted a lively debate between French and German artists in Rome about the appropriateness of introducing such historical veracity into biblical subjects. This debate was recorded by many visitors to the city, including the Scottish history painter David Scott.[10]

Vernet's interest in historical veracity soon was to lead him to take an even more radical step in the portrayal of biblical scenes. In the wake of the French incursions into North Africa he made a series of visits to the Arab world in the 1830s. His principal concern was to gather material for the increasingly popular genre of scenes of Arab life. But when observing the customs of the Bedouins he also became struck with the similarities that must exist between such nomads and the Jewish tribes of the Old Testament. He became convinced that the observation of the life of the Bedouins could provide details relating to biblical sub-

Figure 2.3. Horace Vernet (France, 1789–1863), *Judith and Holofernes*, 1831, oil. Musee des Beaux-Arts, Pau, France

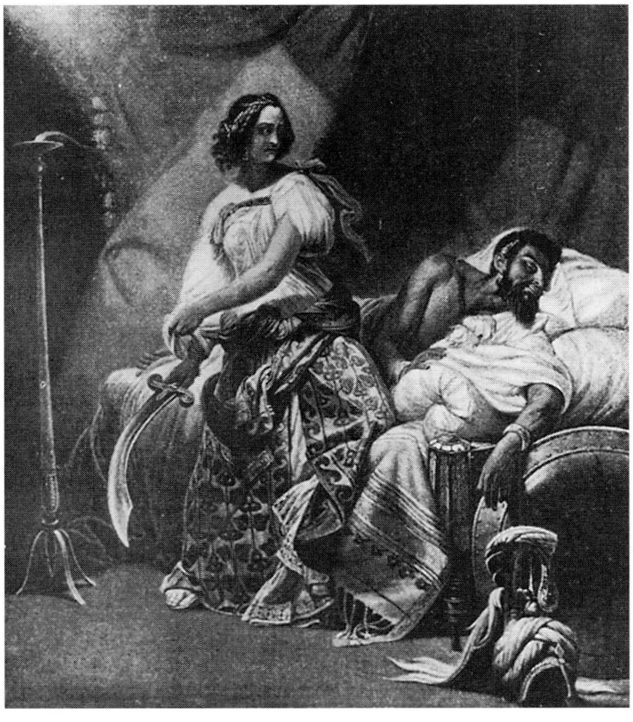

Figure 2.4. After Horace Vernet, *Rebecca and Eliezer*, ca. 1840, oil, original now missing.

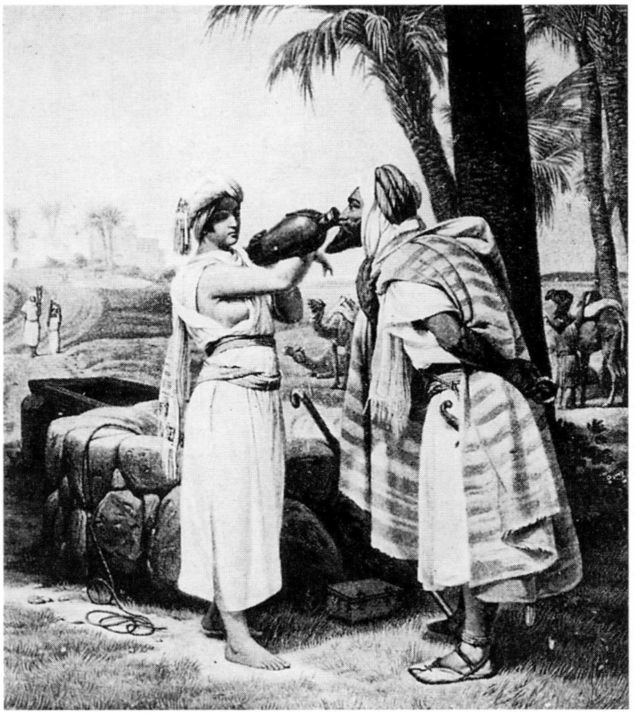

jects that could not be found in any other way. One example of this was provided by the picture that he subsequently painted illustrating the passage in Genesis (24:18) that describes how Rebecca "let down her pitcher upon her hand" to give a drink to Eliezer, the servant of Abraham (fig. 2.4). Later, in a report to the French Institute of 1848 Vernet recounted how he had been reading this passage while camping in the desert in Egypt.[11] He had been puzzled about the detail of how Rebecca actually lowered the pitcher on her hand. He raised his head and saw before him a Bedouin girl offering a drink in a manner that seemed to fit the biblical account exactly. It is this singular gesture that is the focus of his picture. Excited by his discovery, Vernet went on to portray biblical scenes completely in terms of the manners and costumes of the Arab life he had witnessed. He also attempted an accuracy of location. When painting *The Good Samaritan*, for example, he painted the background on the road from Jericho to Jerusalem. Vernet was not, of course, the first artist to introduce Palestinian locations into biblical subjects; one can find such attempts, for example, in the work of the sixteenth-century Flemish painter Jan van Scorel.[12] But Vernet was the first to render them totally in terms of the manners and details of the Near East.

Vernet's reasons for straying into portraying biblical subjects in this manner may have been relatively superficial. It probably arose from a desire to capitalize on his specialist knowledge of the Arab world. It is noticeable, too, that he steered away from all subjects that might have provoked the charge of blasphemy. Most scenes were taken from the Old Testament, and he never depicted Christ. Nor did anyone think of the pictures that he painted as aids to devotion or moral instruction. Even his supporters felt that he had missed the mark from this point of view. As his biographer A. Durand said, his figures "resemble simple Bedouins too much for one to be able to recognize and respect in them the dignity of history."[13]

Vernet's innovation, indeed, would probably not have been taken seriously by religious painters had it not coincided with the new wave of biblical scholarship that sought to establish a historical view of Christ and his times. Vernet probably did not read Strauss's *Life of Christ*, although it is worth pointing out that this book appeared three years before the occasion that led the artist to portray *Rebecca and Eliezer*. But both biblical scholar and painter were profiting from the closer knowledge of the Near East that had emerged as this area became more accessible to Europeans in the 1830s. Vernet was far from being the only painter to visit those parts of the world at the time. Delacroix's journey to North Africa in 1832 is well known. The Holy Land itself was visited in the 1830s by many topographers, notably David Roberts in 1838, whose subsequent lithographs and panoramas made the public in Britain increasingly aware of what the terrain of the locations of biblical scenes actually looked like.[14] Already by 1840 the genre and history painter David Wilkie had been inspired by such views to travel to Palestine to gather material for religious subjects.[15]

Little can be said about what Wilkie would have made of his material; he died on his way back from Palestine. It was a younger generation of artists who were to take up the challenge of representing biblical scenes in a historical manner.

In Britain in the 1840s, the representation of religious scenes became a matter of renewed importance. This was both because of the new stimulus towards painting serious historical subjects evidenced in such events as the competitions between 1843 and 1846 for painting murals for the New Houses of Parliament and because of the new incentive for religious art provided by the revival of catholic ritual in both the Roman and Anglican churches.[16]

Such problems can be seen most clearly in the work of William Dyce, the

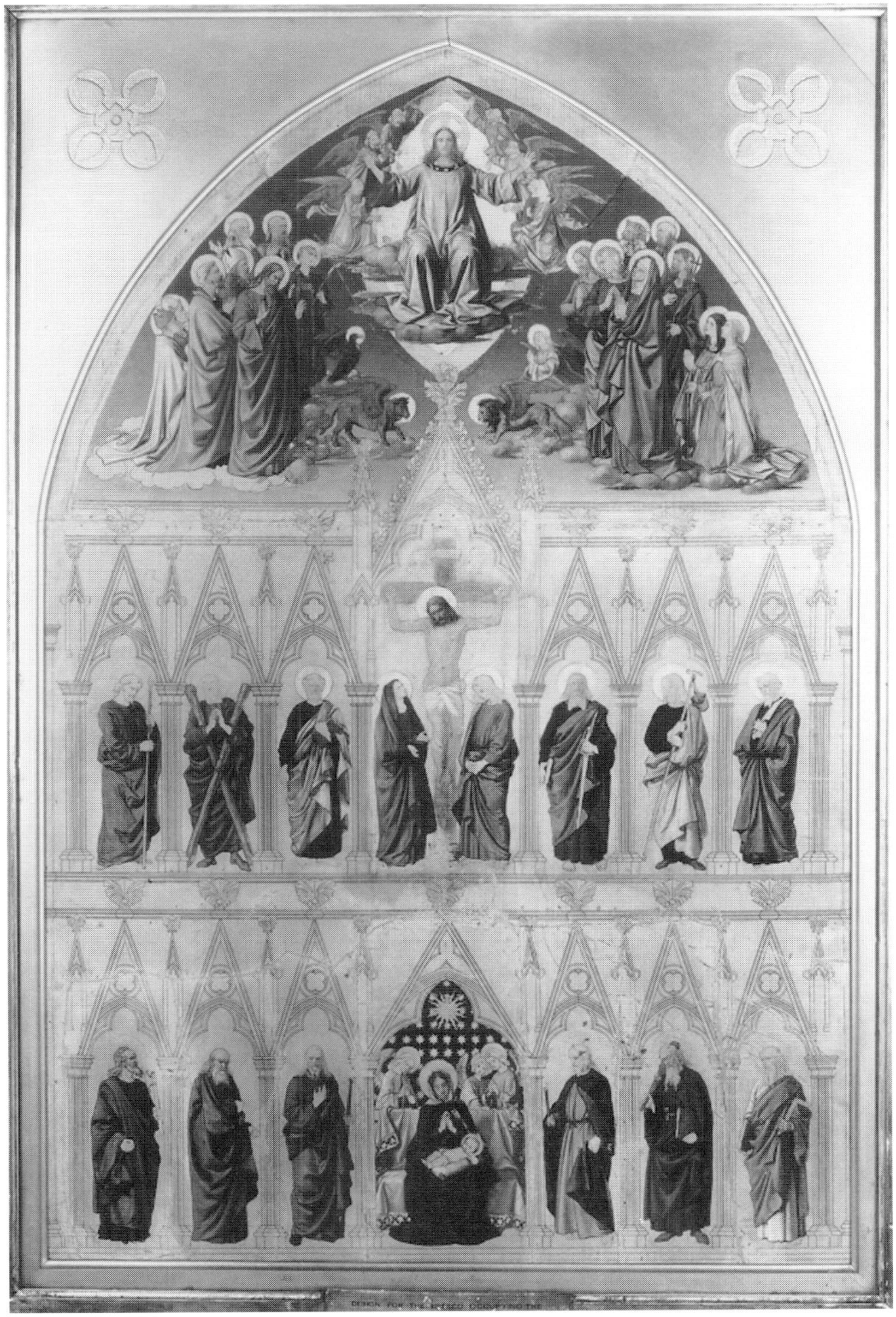

Figure 2.5. William Dyce (Scotland, 1806–1864), *Study for Reredos of All Saints, Margaret Street*, 1849, oil. Victoria and Albert Museum, London

Scottish painter who was a leading figure in both the Houses of Parliament mural scheme and in the projects for church decoration being brought in by high church Anglicans. Dyce had long harbored the desire to be a religious artist, an ambition that had been fanned by his contacts with Overbeck and Wiseman in Rome in the late 1820s. His attempts to establish himself as a painter of religious subjects had met with little success when he first returned to Britain in 1830. But in the 1840s he found opportunities for such work increasing, especially when he won the support of such leading figures in the Anglo-Catholic movement as Gladstone and Beresford Hope. It was the latter who provided Dyce with the opportunity of painting the reredos of the Anglo-Catholic's "show church," All Saints Margaret Street, London, in 1849.[17]

It has often been assumed that Dyce was an English disciple of the Nazarenes. Certainly this painting—to judge from the extant study for it (fig. 2.5)—suggests that he shared their concerns both to revive a quattrocento manner and to reintroduce medieval iconography into religious art. Yet the situation is not quite as straightforward as that. Despite his close contacts with Overbeck and Wiseman, which continued after his return to Britain, Dyce had strong reservations, both artistically and theologically, about the Catholic circle in which he had moved in Rome. He was acutely aware of the distinction between the catholicism of the Anglican Church and that of the Roman. This led him to include certain iconographical variations in his works. In the All Saints Margaret Street reredos, for example, he included the Protestant hero St. Paul amongst the apostles. But more importantly, it colored his whole perception of the function of religious art both inside and outside the Church. He was greatly concerned to avoid any taint of idolatry. On one occasion he wrote with respect to the use of the visual arts in churches, "The abuses to which certain of the arts had ministered it must be remembered, led to their being discountenanced by the Church of England: It is vain, therefore, to look for those which the church abolished, as accessories to devotion."[18] It is hard to see how Dyce's painting for All Saints Margaret Street can avoid being seen as an "accessory to devotion," though it could be argued that, as it was painted on the wall behind the altar it was essentially a wall decoration rather than a devotional work. Equally difficult to absolve from devotional duties are the many Madonnas that Dyce painted. These are the works that seem to come closest to the art of the Nazarenes and are most unequivocable in their emulation of the properties of early Italian art. The one painted for Prince Albert in 1845 was considered "Umbrian" by the press when it was exhibited at the Royal Academy in the following year.[19] It may in fact derive its peculiar profile view from an altarpiece then ascribed to Squarcione, which was reproduced in Mrs. Merrifield's edition of Cennino Cennini.[20] The profile view that is so unusual has led Dr. Robyn Cooper to suggest recently that Dyce avoided a "devotional" aspect to his Madonnas by making them fail to confront the spectator directly.[21] It is certainly true that none of his painted Madonnas look straight out of the picture, although this more conventional gaze can be found in some of his drawings of Madonnas.[22]

EVEN IF ONE pursues such arguments, one cannot help feeling that the line dividing such works by Dyce and traditional devotional works is so thin as to be virtually invisible, however discernible it might have been to the artist. It is undeniable, too, that such works by Dyce, which were often intended for ecclesiastical settings, were based upon traditional devotional types. But there is another kind of scriptural painting that Dyce engaged in which was of quite a different type. This was the biblical subject piece, of which he exhibited a number at the Academy. One of the earliest was *Joash Shooting the Arrow of Deliverance* (fig. 2.6),

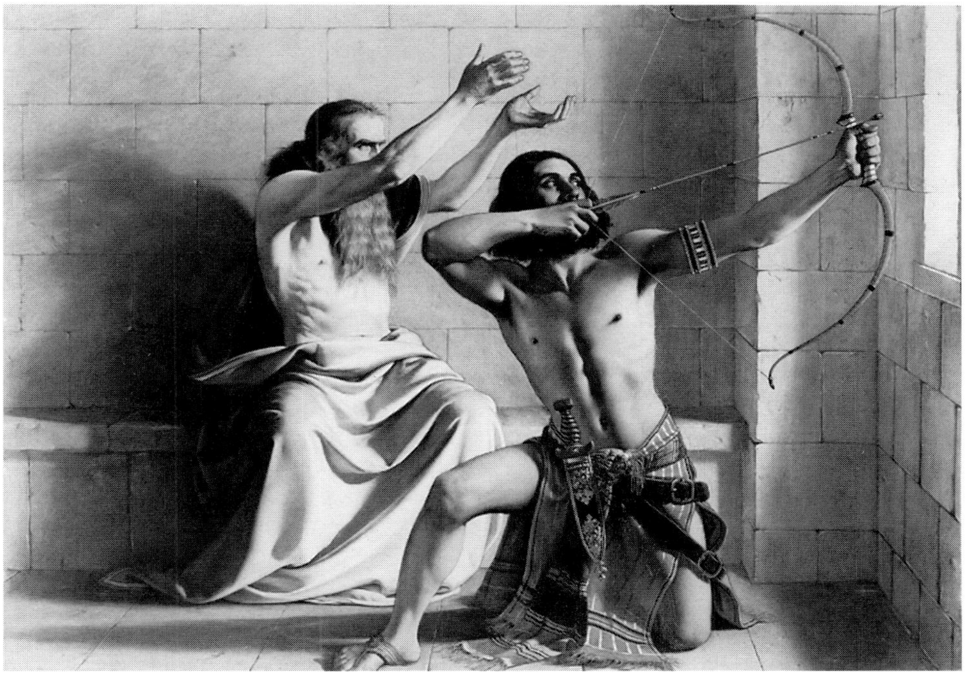

Figure 2.6. William Dyce (Scotland, 1806–1864), *Joash Shooting the Arrow of Deliverance*, 1844, oil. Kunsthalle, Hamburg

an Old Testament theme from the life of Elisha that was exhibited at the Royal Academy in 1844. A comparison with a work painted eight years earlier—*The Judgement of Solomon* (fig. 2.7)—is sufficient to demonstrate how much the rigor of treatment is a development of Dyce's work in the 1840s and not a product of his time in Rome, as some suppose.

But equally striking is the way in which this rigor, which can be related to his renewed contact with German Nazarene painting during his tour of the continent in the late 1830s, is accompanied by a growing concern for historical costume doubtless encouraged by his stay in France. In the place of the vague Renaissance drapes of *The Judgement of Solomon*, *Joash* shows an attempt to reconstruct the costume of pre-Christian Palestine. This feature was in fact the one most commentated upon by reviewers, in terms that suggested they knew of Vernet's innovations. For the German periodical *Kunstblatt* (whose editor, Ernst Foerster, a former pupil of the Nazarene Peter Cornelius, had strong sympathies with revivalism), the advantages of the new procedure were questionable: "I am not sure that the artist has not made a mistake in losing religious and poetic content for the sake of a gain—a so-called true history. For these are of such barrenness that they appear prosaic—as in the case of [Vernet's] *Rebecca* or shocking—as in the case of [Vernet's] *Judith*."[23]

The English magazine *Art Union*, on the other hand, welcomed the treatment, claiming to have "long contended for oriental character as a propriety in scriptural art," and feeling the figure of Joash to reveal "the better part of the Oriental as he now is, and as he was in the days of Solomon."[24]

Dyce evidently encountered a certain amount of opposition from his high church supporters, for he wrote defensively in a letter to his patron Hope Scott about the work, "I fear you were disappointed because you looked for certain characteristics of works by the *ascetic* painters of the early schools—if you did, excuse my saying that you ought to have been disappointed."[25] He then

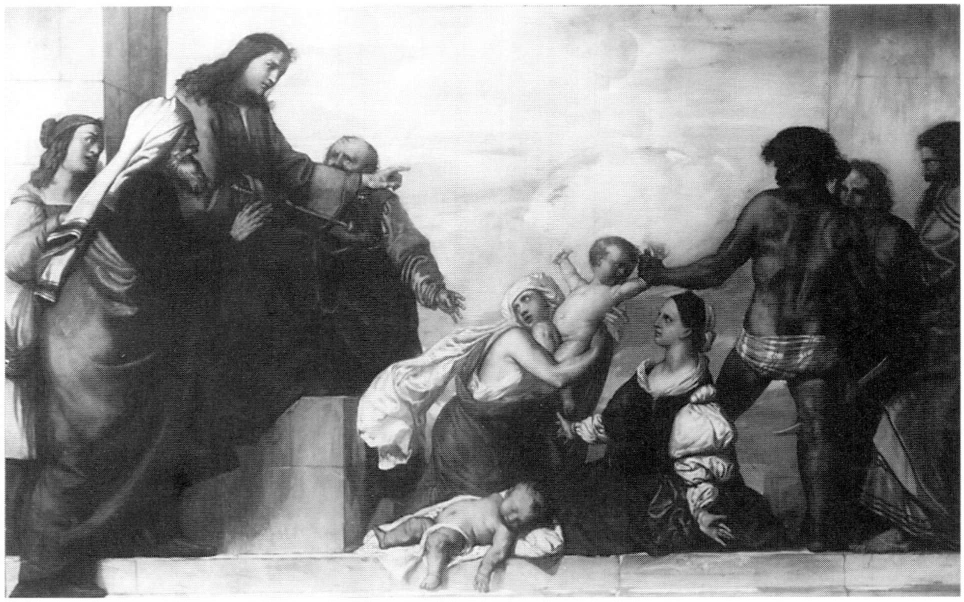

Figure 2.7. William Dyce (Scotland, 1806–1864), *The Judgement of Solomon*, 1836, oil. National Gallery of Scotland, Edinburgh

explained the distinction between secular and sacred representations of themes, emphasizing that the picture in question belonged to the former category and should be judged accordingly. And, in case there should be any objection to a painter of religious subjects producing (as he had done in the case of *Joash*) a "secular" Academy piece, he suggested a symbolic rationale for the work: "The picture is not a devotional kind, I admit; but to those who look beyond the surface it may serve to symbolize the arm of secular power directed by the church."

It is indeed possible to interpret Dyce's unusual and erudite theme in this light. The representation of the descendant of David shooting the arrow of deliverance at the behest of Elisha for the defeat of Aram certainly shows a deference of secular to ecclesiastic authority. As one firmly in support of the "established" Church in Britain, such a theme would no doubt have been especially meaningful to Dyce, as it would have been to his supporters.

The historicism that Dyce developed in his panel pictures of religious and biblical subjects was accompanied by a growing naturalism. Eventually, this naturalistic interest was to dominate and superceded historicist interests. Thus, as Allen Staley has demonstrated, a development can be traced from the quasi-Palestinian setting of *Joash* to such works as *The Man of Sorrows*, which are set in British scenery that the artist knew at first hand. Undoubtedly, as Staley suggests, this development was intensified by Dyce's contact with the Pre-Raphaelites. Amongst that group William Holman Hunt, at least, appreciated the *plein air* element to be found in Dyce's painting. When considering Dyce's relationship to the artists of the past Holman Hunt wrote, "He was not as many modern painters have been, a mere plagiarist of their postures and expressions; in his work could always be seen some sweet trait from the freshness of the passing day over and above the culture of the great masters whose living representative he made himself."[26]

Hunt had personal reasons to be grateful to Dyce. In 1850, when Hunt's fortunes were low, Dyce engaged him to make a copy of one of his biblical subjects, *Jacob and Rachel* (fig. 1.21). This picture certainly combines the "freshness of the passing day" with the "culture of the great masters." But the culture of the

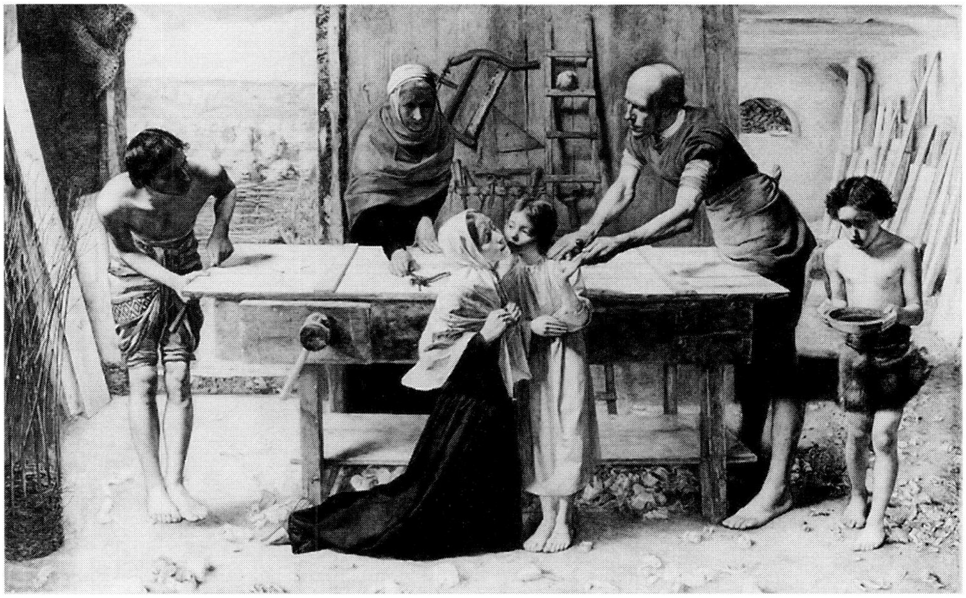

Figure 2.8. John Everett Millais (England, 1829–1896), *Christ in the House of his Parents*, 1850, oil. The Tate Gallery, London

great masters appears to have been absorbed in this case via a Nazarene source. For it appears to be based on a composition of the same subject by the Nazarene Julius Schnorr von Carolsfeld.[27]

Hunt's contact with Dyce at this juncture was highly critical. It occurred at a time when the Pre-Raphaelite fortunes had been almost totally wrecked by the savage reception of their works at the Royal Academy. It had been a religious work that had drawn the most damning comments—Millais' *Christ in the House of his Parents* (fig. 2.8). Like so many religious pictures of the 1840s, this work had attempted to combine a new sense of realism with a devout and spiritual air. Yet in its case the extreme of using non-idealized figures had seemed to verge on blasphemy. It has been suggested that part of the reason for the attack on this work was its association with high church and Catholic sympathies at a time when a violent anti-Catholic reaction was setting in.[28] This may have provided a focus for the hostility felt towards the picture, but it is worth remembering that the picture did not resemble other "Catholic" or "Anglo-Catholic" works of the period. Furthermore, many other Catholic and high church artists—such as Dyce, William C. T. Dobson and J. R. Herbert—exhibited without drawing similar criticism. It was the shocking realism of the work, the "hideous" emaciation of the figures that was at the heart of the objection. Millais' own connections with the Anglo-Catholic movement were relatively casual, and while his picture clearly shows a "tractarian" influence in its use of symbolism this is employed in a highly idiosyncratic manner. There is little to suggest that this work was seen by the Anglo-Catholic group as representing their views. It is significant, too, that the picture had no sequel.[29]

As I have already remarked, Millais' attempt at bringing greater reality to scriptural subjects had many immediate forebears. Like Dyce he attempted to give a spiritual dimension to such realism by allying it with a rigorous design. It may be no more than a coincidence that the only connection between Millais' depiction of this theme of an incident during Christ's apprenticeship in Joseph's workshop prefiguring the crucifixion and a similar subject by Overbeck (fig. 2.9) is the symmetrical pattern adopted by the figures.[30] Yet it may be more than a coincidence, since the Overbeck painting was published in England two years before the painting of *Christ in the House of his Parents* (fig. 42). Certainly one can see that Millais retained an interest in Nazarene art during this period—for he made a memory sketch of a *Madonna* by the Nazarene follower Steinle in the summer of 1850 (fig. 2.10).

Millais did not follow up his radical incursion into religious art. Hunt, on the other hand, experienced a strong religious conversion around 1851 and subsequently made the painting of religious pictures the mainspring of his career.[31] While he remained devoted to the Pre-Raphaelite tenet of literal and painstaking pictorial realism, he had been warned off the pursuit of a too meager realism in religious art by the hostile reception to Millais' *Christ in the House of his Parents*. It was necessary to show reality in a more heroic and healthy guise, and for this Dyce's religious art provided a clue. As Allen Staley has pointed out, of the work by Dyce that Hunt copied in 1850, *Jacob and Rachel* seems to have provided the starting point for the painting that Hunt began immediately afterwards, *The Hireling Shepherd* (fig. 2.11), in both its simple inverted v-shaped meeting of the two principal figures and in its pastoral setting.[32] The pictures are also alike in their use of well rounded, healthy looking figures. Unlike Dyce's work, Hunt's is basically allegorical. The shepherd and shepherdess who are dallying while their flock go astray symbolize the "muddle-headed pastors" of the time who engaged in arid theological dispute while the souls in their care go astray. In fact, Hunt's

Figure 2.9. Johann Freidrich Overbeck (Germany, 1789–1869), engraved by X. Steifensand, *Puer Jesus in Fabrica Josephi*, engraving, plate 9 in *Die Vierzig Evangelischen Darstellungen* (Düsseldorf, 1847).

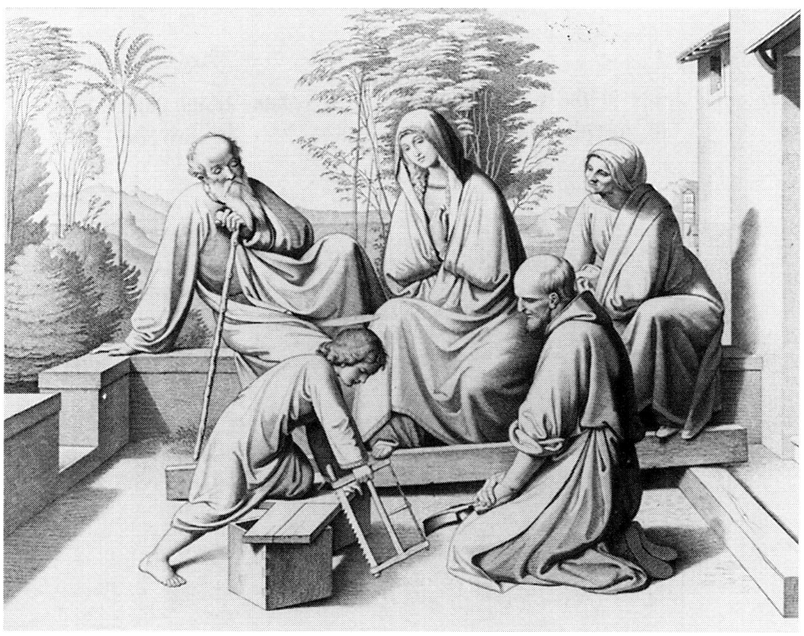

Figure 2.10. John Everett Millais (England, 1829–1896), *Memory Drawing of Madonna by Steinle*, 1850, pencil. Department of Prints and Drawings, Ashmolean Museum, Oxford

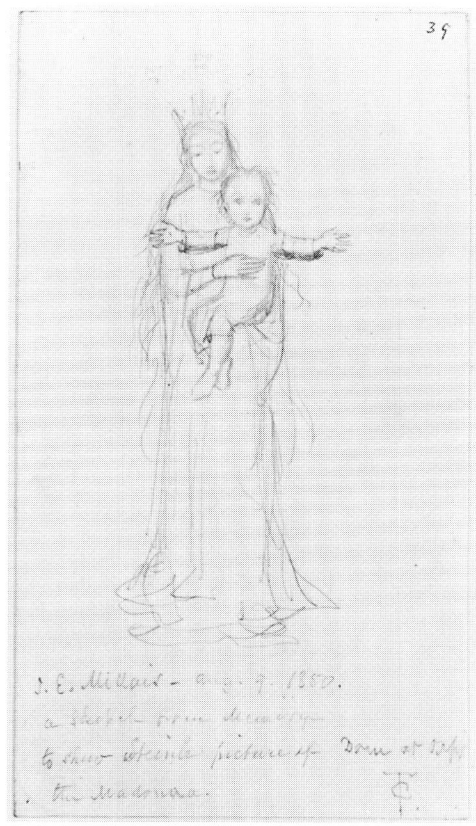

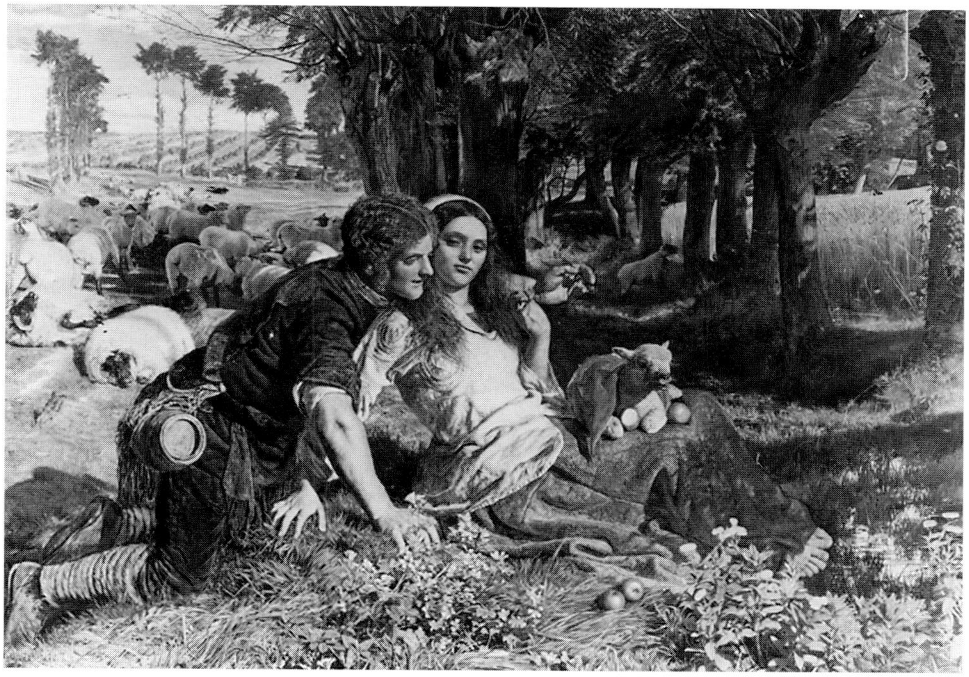

Figure 2.11. William Holman Hunt (England, 1827–1910), *The Hireling Shepherd*, 1851–52, oil. City of Manchester Art Galleries, Manchester, England

commitment to pictorial realism was such by this time that he could paint religious pictures only in allegorical form while he remained in Britain. Unlike Dyce, he could not be content to reconstruct Palestinian settings for "real" biblical subjects without having gone to see the place.

Hunt's decision to go to Palestine in the wake of Vernet, Wilkie, and other artists was a consequence of this predicament. The confused reaction to his religious allegories—neither *Hireling Shepherd* nor *The Light of the World* received a good press when they were exhibited—seemed to convince him that he would never be able to address a wider public unless he moved from the esoteric field of allegory to the direct relation of biblical events. On the other hand, he was well aware of the pitfalls that awaited the artist who attempted a too prosaically realistic rendering of biblical scenes. His fellow Pre-Raphaelite Rossetti had reminded him of the failure of Vernet in this direction and used the example of the French artists to try and prevent his colleague from traveling to the Holy Land.[33] Yet Hunt remained convinced that the rendering of biblical scenes—if approached in the right spirit—could provide compelling contemporary devotional works. Like his mentors Carlyle and Ruskin, he came from an evangelical, low church background and supported the idea of a literal interpretation of the Bible. He therefore believed that there was a historical basis to the scenes of the life of Christ, and that by making this historical fact a pictorial reality he would be enhancing the authority of the Christian message. He therefore felt the need, as Carlyle did, of confronting the new generation of biblical scholarship, not by disputing whether the historical Christ could be found, but by disputing the assertion that this historical Christ was less than the figure presented by the Christian faith. Hunt's problem was to reconstruct the historical Christ in a way that asserted rather than undermined his divine stature.

It was a problem that he found daunting. And while he planned to paint a historical subject during his first visit to Palestine in 1854, *Finding of the Saviour in*

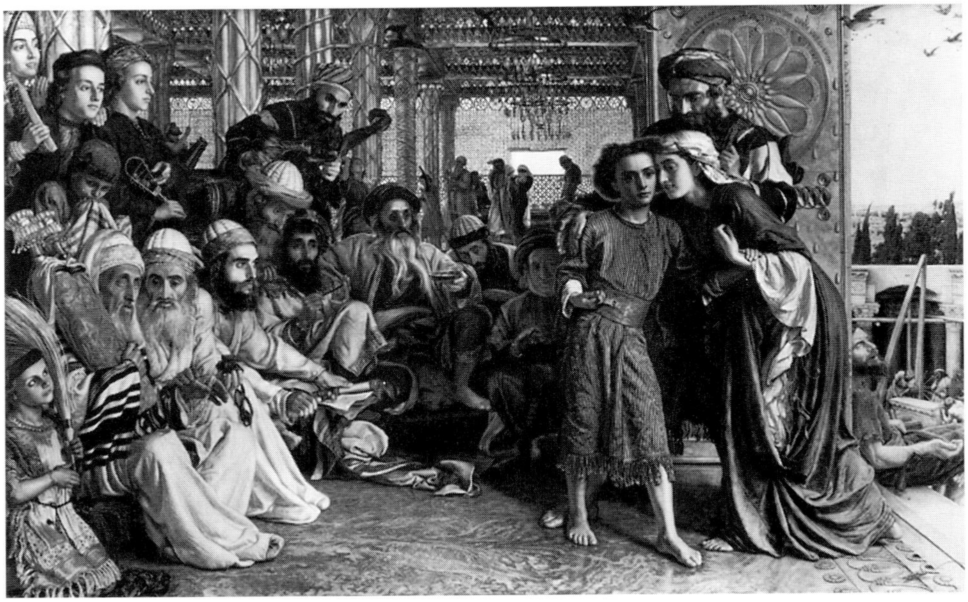

Figure 2.12. William Holman Hunt (England, 1827–1910), *The Finding of the Saviour in the Temple*, 1860, oil. City Museum and Art Gallery, Birmingham, England

the Temple (fig. 2.12), he found the practical problems (which included finding suitable models) overwhelming and fell back instead into his allegorical habits. The reception of the resultant *The Scapegoat* (fig. 2.13) when it was exhibited at the Royal Academy in 1856 added further proof that allegory was not the path through which to achieve popular acclaim. Hunt's depiction of the symbolic bearer of the sins of the Israelites wandering desolately before the Dead Sea did not strike viewers immediately as a prefiguration of Christ in the way that he had intended. Even Ruskin—who had defended his earlier allegories—disapproved.

When *Finding of the Saviour in the Temple* was finally completed, it proved to be a triumphant vindication of the historical approach. The work was shown at a special exhibition at the gallery of the dealer Gambart in 1860. Gambart paid Hunt the then huge sum of £5,000 for the picture plus the copyright. This contrasted with the £600 with copyright that Hunt had received for the *The Light of the World* a few years earlier.[34]

Part of the reason for Hunt's success was the undoubted interest that the topographical accuracy of the work afforded. But equally important was the fact that this realistic rendering of a scene from the life of Christ did not in any way offend traditional notions of the appearance and dignity of Christ. Hunt went to great lengths to find a suitably glamorous model for the young Christ, and the trouble paid off. The *Athenaeum* reviewer remarked, "We can conceive nothing could be more successful than this face...the refinement, beauty and dignity surrounding this personification of our Saviour are only enhanced by the pathos which the action of the hands suggest."[35]

The picture provided the most poignant contrast to the work by Millais that had been so ill received ten years earlier. Both showed scenes from Christ's childhood that presaged his future. But whereas Millais' subject was utterly fictional—a romance on an apocryphal tradition elaborated by the artist—Hunt's was one that was universally considered to be a historical fact. Even Strauss considered the story of Christ in the Temple to be based on a real event—contrasting it in this respect with the mythical nature of the miracle stories.[36] And whereas

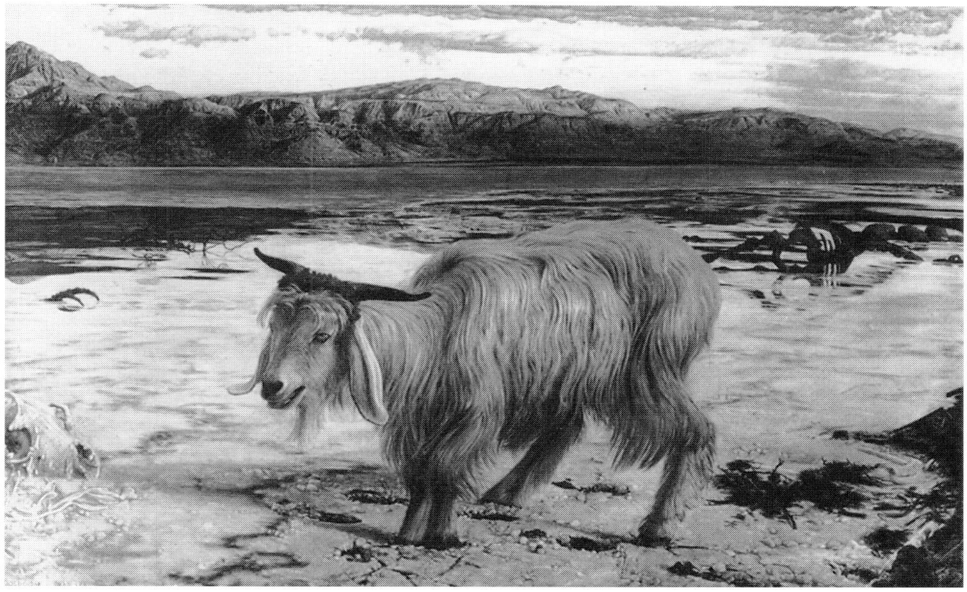

Figure 2.13. William Holman Hunt (England, 1827–1910), *The Scapegoat*, 1856, oil. Lady Lever Art Gallery, Port Sunlight, England

Millais' fictional subject was peopled with figures who were not idealized (and who were chosen largely from amongst his family and friends), Hunt made use of a carefully selected "superior" physical type to emphasize the divine dimension within the historical event. This last point is perhaps the most significant of all. For by representing Christ in a manner that seemed to contemporaries to suggest health and beauty, he was interpreting the story in terms of an "agreeable reality" that enhanced rather than conflicted with traditional beliefs.

Hunt's most acclaimed work in this genre was completed some thirteen years later. Titled *The Shadow of Death* (fig. 2.14), it was also a subject of prefiguration, showing an accidental shadow cast by Christ while working in the carpenter's shop that presaged his crucifixion. This time the subject was a fictional one, but Hunt went to great lengths to attain topographical and historical exactitude, painting the subject in a carpenter's shop in Bethlehem and taking care to include only objects that he believed to have been in existence at the time of Christ. The figures, too, were based on local types. Once again he sought to convey the divinity of Christ through his physical presence. Rather interestingly this fact was seen by one reviewer, J. B. Atkinson, to be in line with the views of one of the rationalist interpreters of the Bible. This time the interpreter was not Strauss but Joseph Ernst Renan, whose *Vie de Jesus* had appeared in 1863: "The figure, though lacking dignity, realizes at least the rationalistic idea of M. Renan, that the spell of Christ's personal presence proves a physical beauty and perfection approaching the superhuman."[37]

Unlike Strauss, Renan had gone beyond attempting to sort out the mythical from the historical in the account of Christ's life and had developed a positive assessment of his character. His book was, in fact, more like a biography and attained all the popular success that is so frequently accorded to that genre. Yet despite the fact that it kept less closely to the evidence than did Strauss's study, it was still seen as a "rationalistic" work by virtue of its attempt to account for the life of Christ in purely material terms. Renan's view that Christ's physical beauty must have been extreme for him to command the attention he did is typical of his approach.

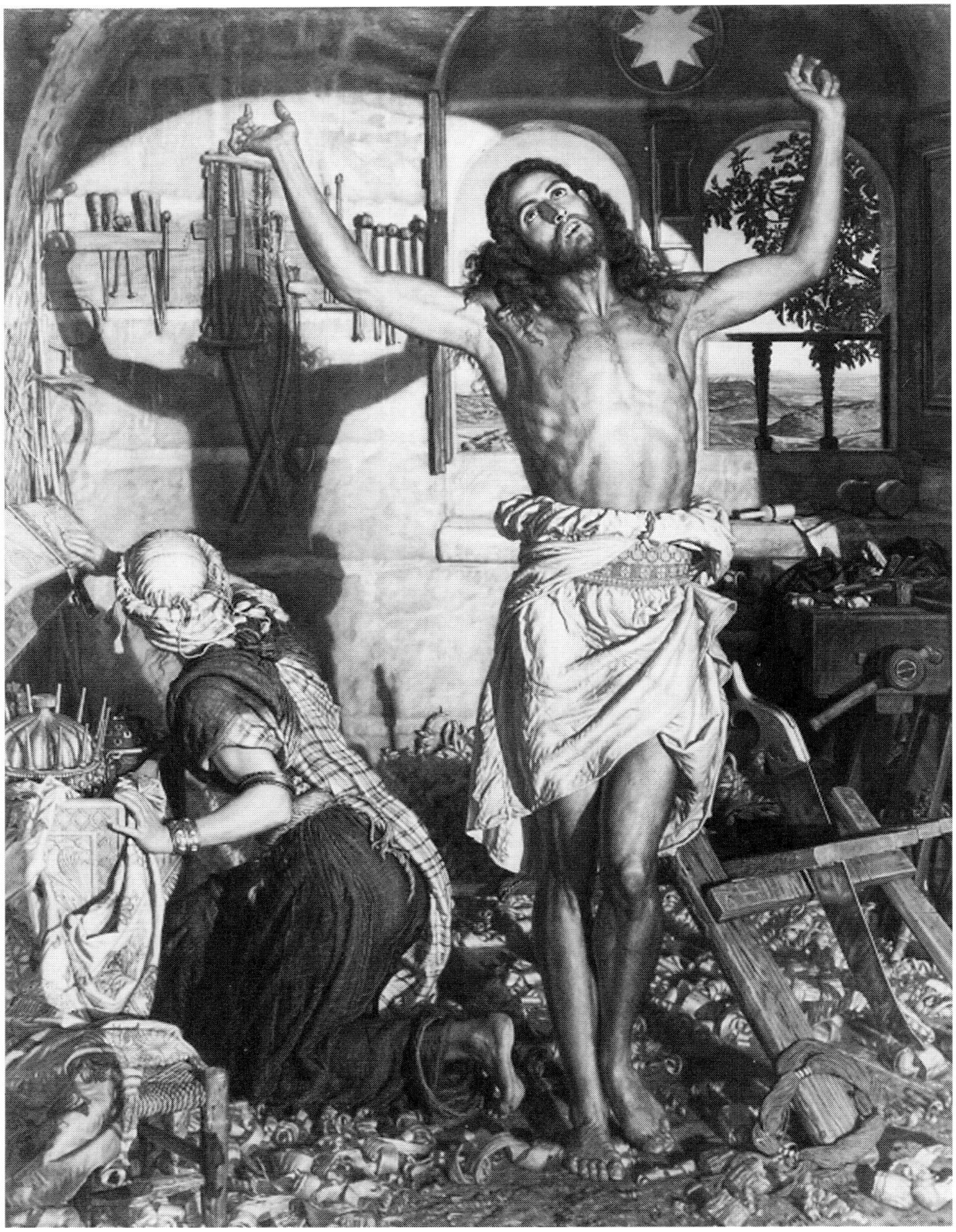

Figure 2.14. William Holman Hunt (England, 1827–1910), *The Shadow of Death*, 1870–73, oil. City Art Gallery, Manchester, England

It would be interesting to know if Hunt had read Renan's account of Christ's life before the time that he painted *The Shadow of Death*. He certainly had a decade later in 1883, when he wrote of it in disapproving terms to William Bell Scott, "Renan's works I have read exhaustively; they are, in spite of scholarships which make them worth reading, simply puerile nonsense with about as much knowledge of his subject as Payne Knight had of the Elgin Marbles."[38]

Hunt objected—as well he might as a Christian with evangelical sympathies—to the "critical contempt of the evidence on which revealed religion stands." Yet his distinction between Renan's knowledge and his scholarship is a telling one. He evidently found Renan's detailed investigations of his subject stimulating, even if he felt that the French author lacked the knowledge—the understanding afforded by religious belief—to read his evidence aright.

Seen from this point of view, *The Shadow of Death* provides an interesting counterpoint to Renan's account of Christ's early years.[39] Renan seeks to maintain that Christ was born at Nazareth, rather than at Bethlehem, and dwells lyrically on the people and scenery of the former. Hunt went to Bethlehem to paint his scene from Christ's life. Renan felt that Christ was a man of the people and rejected the claim that he was descended from the line of David. Hunt emphatically includes in his picture the star of David, incorporating it into the window frame at the back of the room. He also brought in references to other incidents in the Gospels that Renan—like Strauss—considered to be mythical, such as the gifts from the three kings that Mary is placing in the chest on the left of the painting.

But if Hunt and Renan are at odds on those details that touch on Christ's divinity, there is some similarity in their description of the day-to-day details of Christ's life. Both artist and writer looked to contemporary shops in Bethlehem and Nazareth to suggest the location where Christ lived.[40] As Renan said, "The house of Joseph doubtless much resembled those poor shops, lighted by the door...having for furniture a mat, some cushions on the ground, one or two clay pots, and a painted chest."[41]

But more interesting than this was the emphasis that both artist and writer placed on Christ's proletarian upbringing. Renan's view of Christ as a laborer was notorious and had already led to his name being associated with realist religious art in France. As early as 1864 Manet's *Dead Christ with Angels* was satirically described in *La Vie Parisienne* as "'The Poor Miner Pulled out of the Coal Mine' painted for Renan."[42] Renan described Christ's parents as "craftsmen, living by their labor, in a state so common in the East which is neither ease nor poverty." Hunt deliberately showed Christ as stretching after a day of hard but invigorating labor. For the artist it was a particularly important point to emphasize this side of Christ, since it enhanced his relevance for the social problems of the nineteenth century. Hunt took pride in the fact that the print of his picture sold well amongst the working classes in the north of England.[43]

Subsequently Hunt's *The Light of the World* has proved a more haunting and enduring image than *The Shadow of Death*. Yet there is no doubt that the latter was more effective at the time. As late as 1886 Ruskin referred to *The Shadow of Death* as being generally considered Hunt's most successful work.[44] Perhaps it appealed so much because it seemed to provide an answer to an urgent need of the age—a need to accept Christ both as a divinity and as a vivid historical reality, a figure who could survive the investigations of the "materialist" historians. Hunt claimed that he had suggested both of these ideas without the one destroying the other. Christ looks superhuman, yet there is no supernatural occurrence in the work. With this picture, as J. B. Atkinson put it, "Materialism acquired a spiritual significance."[45]

Hunt's infusion of the "real" facts about Christ with a visible sense of the spiritual may have been inspired by his evangelical background. Yet his achievement was to have implications for religious art across the denominations. This can be seen by the impact of his approach amongst Catholic artists in France. The shock of Renan's *Vie de Jesus* in 1863 had certainly led to a more careful consideration of historicist detail. This can be seen in the religious work that Gustave Doré began to paint from the late 1860s, such as *Le Christ sortant du pretoire* (1867–72), exhibited with great success both in Paris and London in 1872.[46] Yet there is no sense of direct experience in the work. The details are not based on studies from the Middle East, and the picture is full of references to the Old Masters in both its design and handling. The figure of Christ, moreover, still has a nimbus. Quite different is the position of artist James Tissot, who began to turn to religious subjects after a visit to Jerusalem in 1886. Tissot's series of pictures of biblical scenes—particularly the three illustrations published as *La vie de Notre Seigneur Jesus Christ* in 1894 and called by Edward de Goncourt a "*monographie realiste de Jesus*"[47]—show the adoption of Hunt's practice of representing the life of Christ in terms of the artist's direct perception of the Near East. Like Hunt, too, Tissot sees the "reality" that he has restored here as being different from the "realism" of the materialists.[48] Tissot's religious work has been related to the Catholic revival of the last quarter of the nineteenth century,[49] yet just as Hunt's pictures appealed across the denominations, so did Tissot's. His *Life of Christ* was as popular amongst Protestants in Britain and the United States as it was amongst Catholics in France.

To judge from the accounts of enthusiastic spectators, Hunt and Tissot were highly successful in producing works that could act as aids to devotion. Yet it is also clear that they became distanced thereby from current intellectual and artistic developments. Scriptural scenes of "agreeable reality" have subsequently come under fire from both realists—who would like something more gritty—and idealists—who would like something more transcendent. Only in an age when there was still a notional continuity between rational inquiry and divine revelation could such pictures satisfy the interests of both groups.

Notes

1. Quoted by Walter Armstrong in "Sir John E. Millais," *The Art Annual*, supplement to *Art Journal*, 38:25 (1887).

2. L. Errington, *Social and Religious Themes in English Art, 1840–60* (New York: Garland Press, 1984).

3. George P. Landow, *William Holman Hunt and Typological Symbolism* (New Haven, Conn., and London: Published for the Paul Mellon Centre for Studies in British Art by Yale University Press, 1979), 4–6. Landow develops the concept of typology in relation to Victorian literature and art in general in *Victorian Types: Victorian Shadows. Biblical Typology in Victorian Literature, Art and Thought* (Boston: Routledge & Kegan Paul, 1980).

4. M. Bennett, *William Holman Hunt* (Liverpool, England: Walker Art Gallery, 1969), 33–34.

5. W. H. Hunt, *Pre-Raphaelitism and the Pre-Raphaelite Brotherhood* (New York: Macmillan, 1905), 1:355.

6. D. F. Strauss, *Das Leben Jesu*, 2 vols. (Tübingen: J. Osiander), vol. I.

7. O. Pfleiderer in his introduction to D. F. Strauss, *Life of Jesus* (London: Swan Sonnenschein & Co., 1898), xxvi.

8. N. Wiseman, *Four Lectures on the Offices and Ceremonies of Holy Week* (London, 1839), 8.

9. Charles Eastlake, "German Paintings in Rome," a review of Casa Bartholdi frescoes, *London Magazine* (London, 1820), 2:42.

10. W. B. Scott, *David Scott* (Edinburgh, Scotland, 1850), 111.

11. J. Ruuz Rees, *Horace Vernet* (Paris: Didot, 1880), 22ff.

12. For example in the Lochorst Triptych, ca. 1525–27 (Centraalmuseum, Utrecht, Netherlands). See G. von der Osten and H. Vey, *Painting and Sculpture in Germany and the Netherlands: 1500–1600* (Harmondsworth, [Middlesex]: Penguin Books, 1969), 182.

13. A. Durande, *Joseph, Carle et Horace Vernet* (Paris: Hetzel, 1863), 332.

14. K. Sim, *David Roberts R.A.* (London: Quartet Books, 1984), 220. Roberts' major work in this area was *The Destruction of Jerusalem under Titus, A.D. 71*, exhibited at the Royal Academy in 1850; see Sim, 253–55).

15. A. Cunningham, *The Life of Sir David Wilkie* (London: J. Murray, 1843), 939ff.

16. See W. Vaughan, *German Romanticism and English Art* (New Haven, Conn.: Yale University Press, 1979), especially chap. 7, "Dyce and Ecclesiastical Art," 227ff.

17. M. Pointon, *William Dyce* (Oxford: Clarendon, 1979), 128ff.

18. W. Dyce, "On Ecclesiastical Architecture: A Defence of Anglican Usage," *Dyce Papers* (Aberdeen Art Gallery, 184[illegible]), sec. 13, 515. See Vaughan, 239.

19. *Illustrated London News* (1845), 337.

20. Mrs. Mary Merrifield, trans., *A Treatise on Painting Written by Cennino Cennini in the Year 1437…* (London: Edward Lumley, 1844), pl. 9, opp. 159.

21. In conversation with the author.

22. See for example *Madonna and Child*, 1848, silverpoint. Department of Prints and Drawings, National Gallery of Scotland, Edinburgh (*Dyce Centenary Exhibition*, Aberdeen Art Gallery, 1964, no. 62).

23. *Kunstblatt* (1844), 293.

24. *Art Union* (1844), 158.

25. Dyce to Hope-Scott, 12 June 1844, Ms. 3669 National Library of Scotland, Edinburgh.

26. Hunt, *Pre-Raphaelitism*, 35–36.

27. See Vaughan, *German Romanticism and English Art*, pl. 120.

28. Errington, *Social and Religious Themes in English Art*, p. 204.

29. Vaughan, *German Romanticism and English Art*, 238–39.

30. *Puer Jesus in Fabrica Josephi*, engraving after Overbeck, *Die Vierzig Evangelischen Darstellungen* (Düsseldorf, 1847), pl. 9. (Reproduced in Vaughan, plate 155).

31. Landow, *William Holman Hunt and Typological Symbolism*, 37.

32. Frederick J. Cummins and Allen Staley, *Romantic Art in Britain* (Philadelphia: Falcon Press, 1968), 273.

33. According to Hunt, Rossetti objected: "... that attention to chronological costume, to the types of the different races of men, to climatic features or influences, were of no value in a painter's work; and that therefore Oriental properties in the treatment of Scriptural subjects were calculated to destroy the poetic nature of a design. He would instance Horace Vernet's picture, painted in the East *Rebecca giving Eleazer to Drink* and some others, as proof of the correctness of his opinion. But I used to meet this by insisting that Vernet, though a remarkably skillful composer and executant, was not, and could not, under any conditions or system, be anything but dull, except to the dull, and was altogether destitute of every spark of poetic fire."W. H. Hunt, "The Pre-Rapahelite Brotherhood—A Fight for Art," pt. 2, *Contemporary Review* 49 (Jan.–June 1886), 740.

34. J. Maas, *Gambart, Prince of the Victorian Art World* (London: Barrie & Jenkins, 1975), 114–22.

35. Bennett, *William Holman Hunt*, 40.

36. D. F. Strauss, *The Life of Christ, Critically Examined*, trans. Evans (1846). See especially the introduction, "Development of the Mythical Point of View in Relation to the Gospel Histories," 1ff.

37. J. B. Atkinson, "William Holman Hunt," *Blackwood's Magazine*, 139:547 (1886).

38. A. C. Gissing, *William Holman Hunt* (London: Duckworth, 1936), 231.

39. J. E. Renan, trans. by C.E. Wilbour, *The Life of Jesus* (New York: Carleton, 1864), 48.

40. Hunt, *Pre-Raphaelitism*, 275.

41. Renan, *Life of Jesus*, 48.

42. See Manet's *Dead Christ with Angels* in the Metropolitan Museum of Art, New York. See also L. Nochlin, *Realism* (New York: Penguin, 1971), 56.

43. Bennett, *William Holman Hunt*, 51.

44. John Ruskin, "Notes to the Exhibition of Hunt's Work," *Fine Arts Society*, 28:37 (1886).

45. Atkinson, "William Holman Hunt," 545.

46. A. Renonciat, *La vie et l'oeuvre de Gustave Doré* (Paris: ACR Edition; [Lausanne]: Bibliothèque desarts, 1983), 168–70.

47. Quoted in M. Wentworth, *James Tissot* (Oxford: Oxford Universtiy Press, 1984), 187. Wentworth makes no comment on Tissot's debt to Hunt.

48. J. Tissot, *The Life of Our Saviour Jesus Christ* (London: Sampson Low & Co., 1896), 1:ix.

49. See Ian Thompson, "Tissot as a Religious Artist," in *James Tissot*, ed. K. Matyjaszkiewiez (Oxford: Phaidon, 1984), 92–93.

III. Primitivism and Progress: A Victorian Problem?

The Primitive Within

THE CONCEPT OF the "primitive" is encountered in numerous societies. Over the last two centuries, however, it has played a particularly prominent role in the "western" cultures of Europe and North America. Definitions, during this period, have been multifarious and have ranged in tone from contemptuous to wildly adulatory. In the creative arts, in particular, the tendency has been toward the latter; the celebration of a "primitive" aesthetic has been a key feature of the avant-garde. Essentially it is one of the prime validations of "modern art" that it has been able to encapsulate a "primitive" element in its structure and return to western society a seemingly lost perceptual process. This, essentially, is the argument that one finds in such classic studies as Robert Goldwater's *Primitivism in Modern Painting*, which first appeared in 1938, and that was still being asserted in the major exhibition *Primitivism in Twentieth Century Art* held at the Museum of Modern Art, New York, in 1984.[1] One of the works most frequently cited by upholders of this point of view is Picasso's

Figure 3.1. Pablo Picasso (Spain, 1881–1973), *Les Desmoiselles d'Avignon*, 1907, oil. Museum of Modern Art, New York

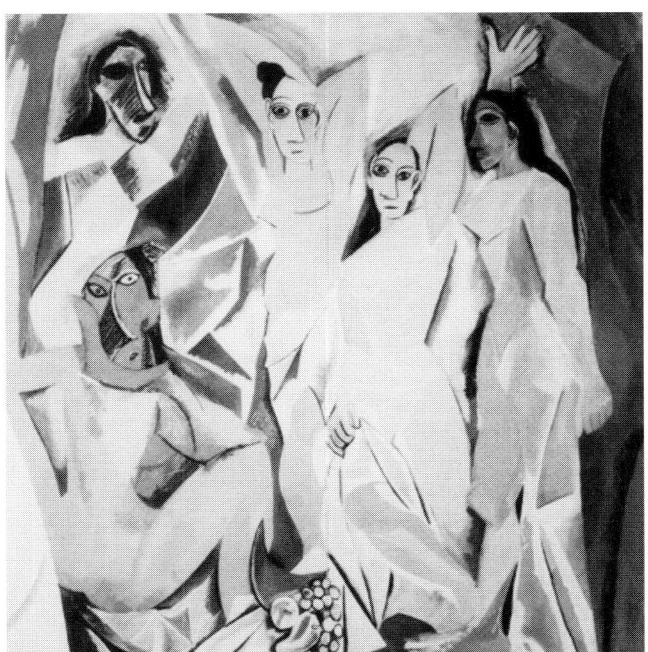

Desmoiselles D'Avignon of 1907 (fig. 3.1). This painting is seen to have derived much of its energy and exciting exploration of space and expressive form from the assimilation of the characteristics of certain African masks that Picasso had encountered in Paris.

No one, of course, would seek to doubt that Picasso had derived inspiration from such sources. What is in question is the role the concept "primitive" plays in this transaction. Until quite recently most art critics and historians would have been happy to describe the African mask as *being* primitive and the response evident in Picasso's picture as being an assimilation of certain of its "primitive" values. But how valid is such an objective use of the term?

Over the past few decades those who have sought to study such art forms in terms of the cultures that produced them increasingly have found that the concept of primitivism is both misleading and demeaning. The idea that categories of non-European art—and of so-called European "folk" art—express some universal "primal" consciousness ignores the fact that such works were made with specific functional and symbolic purposes. It judges works by standards alien to the societies for which the objects were made. As the anthropologist Nelson H. Graburn has put it in his introduction to the collection of studies, *Ethnic and Tourist Arts*, which he edited in 1976:

> Though the terms "primitive" and "folk" art may have been satisfactory for the purpose of nineteenth-century Europeans, it now seems clear that such categories are hopelessly inadequate for any contemporary description. Indeed, they are often taken to be either prejudicial or patronizing slurs upon the arts and artists in question. Even the label "art" itself reflects the traditions of "high civilizations" concerning the value of arts versus crafts, the importance of creativity and originality, and specializations and distinctions that emerged in Europe and China.[2]

Certainly it is clear that the term "primitive" is an imposed definition. No society, I imagine, has ever described *itself* as "primitive" or, if it has done so, it can surely only have been as a consequence of accepting the valuation of another society that had influence over it.

The anthropologists are surely right to make us aware of how arrogant and insulting it is for us to describe cultures alien to our own as "primitive." But does that mean that the concept itself, the notion of the "primitive" as a state of being, can be abandoned? I think not. For just as it is clear that "primitive" has no objective value as a term to describe other cultures, so it is clear that *within* a given culture the notion of the "primitive" is a necessary term of reference, even if this is only to locate a fictionalized "point of departure" for the setting up of the society itself.

One can go further than this and suggest that the notion of the "primitive" is necessary wherever there is a sense of linear historical development. The idea that a society has progressed or even simply changed necessarily predicates the idea that there is some former state from which it has departed. But how is one to know what that former state was? The present developed condition will be clear to us. It is all around us. The so-called "primitive" state, by contrast, has to be extrapolated or recreated. It is the need to perform this reconstruction, in fact, that fuels the quest for signs of the "primitive" in the contemporary world. The so-called developed society is like a wealthy orphan desperately seeking its parents and trying to discover news of them from every passerby.

Primitivism, therefore, is a "problem" created by progress. In the subtitle to this chapter I have called it a "Victorian" problem. This is because the particular issues I wish to explore came into prominence in the mid-nineteenth century. I do not wish to imply that the relation of primitivism to progress is *only* a

Victorian problem. It is every bit as much with us today. I hope the version of it that is described in this chapter will have some bearing on the predicament that exists in the contemporary art world.

In anticipation of this, I would like to make a few further general points about the concept of "primitivism" before proceeding with my particular subject.

The first is that it is used in a "relational" rather than an "absolute" sense. The concept of the primitive may be necessary to any society that has a sense of its own development. But it can only be perceived when there is some other group to which that society can relate itself. This relation can be established in one of two dimensions. These dimensions were characterized by Lovejoy and Boas in their classic study of forms of primitivism published in 1935.[3] These authors referred to them as the "chronological" and "cultural" concepts of the primitive. The "chronological"—or perhaps one might say "historical"—dimension is the one by which a society relates to its own past. Possibly this is always the first instance in which the sense of the primitive emerges. Certainly in the history of western art over the last two centuries the interest in the "primitive" emerges first in the reevaluation of the art of the ancient Greeks and, a little later, of the European Middle Ages.

This is the type of concept of the primitive that is active in a picture discussed in Chapter Two, namely *Christ in the House of his Parents* (fig. 2.8), which was painted by the Pre-Raphaelite artist John Everett Millais in 1850. In this work Millais sought—amongst other things—to capture some of the virtues of the so-called Italian "primitives," that is, the painters who worked before the age of Raphael. I will leave any commentary on the actions taken by the artist to put this into effect until later in this chapter.[4]

The second or "cultural" dimension of primitivism is the one that is brought into being by the difference between one contemporary society or group and another. Often there is a geographical distancing, as in the case of the modern European artists' responses to non-European cultures. But "cultural" primitivism can be generated by internal as well as external distinctions. This occurs when one group in a society perceives the "primitive" in the distinction between itself and another section of that society, as, for example, the urban bourgeoisie looking at a remote rural region. Examples of this kind of "primitive" also abound in the history of art of the nineteenth and twentieth centuries, as in Gustave Courbet's use of the popular print the "Image D'Epinal," or in the "discovery" of naive painters like "le Douanier" Rousseau.

In view of this situation anthropologists are now frequently in the habit of referring to what was once called the "primitive" as a "special relationship" between two groups. This is fine, as long as one remembers that the "special relationship" exists only in the mind of one of the two partners. No group, as I have already said, elects to call *itself* primitive. It is called primitive by the group that perceives itself to be the more advanced. This is perhaps not such a material point for the anthropologists who use the concept, since their principal concern is with the group that is being perceived. Mine, however, as a historian of European art, is essentially with the group that is doing the perceiving. My concern is to gain a closer understanding of how the concept of the "primitive" emerges and how it works.

A more appropriate model for this kind of inquiry might be provided by Edward Said's magnificent study of *Orientalism*.[5] For this work looks at the historical development of a cultural construct created by Europeans to define a multifarious group of non-European communities and takes careful note of the issues as they are forged within the defining society. Yet Said—quite appropriately for the nature of his particular inquiry—is more concerned with the relationship between the cultural construct (orientalism) and the non-European communities it seeks to describe than with the way that cultural construct relates to other concerns within western society. He is not concerned with the interplay between chronological and cultural projections that has constituted such an important part of the elaboration of the concept of the primitive in Europe. Even when he is speculating, for example, about the impact of so-called "scientific" racial theory on the perception of the oriental in the late nineteenth century, he does not address the issues internal to the defining culture that gave such arguments force at that time. It was not, I imagine, necessary for him to do so for the purposes of his investigation. The issue only becomes a problem, in fact, when Said's study is taken—as it has been in recent years—as a model for investigations into the character of primitivism.

For the important point is that "chronological" and "cultural" primitivism cannot be separated. They coexist in the minds of the perceiver. They fuel each others' assumptions. The fact that they are aimed at different "objects" should not obscure the fact that their mode of operation is essentially the same. Indeed, I would go further and say that the distinction between "chronological" and "cultural" primitivism is only one of direction and not of substance. For those who see the "primitive" as an objective entity—as actually residing within the group that is being defined—there is some purpose in making a distinction between "chronological" and "cultural" primitivism. This is, indeed, why an issue of the matter has been made so many times in the last two centuries. But for those who—like myself—see primitivism as a relational term, there is not such a distinction. In considering the issue in relationship to Victorian art and society in this chapter, therefore, I shall be treating the "chronological" and "cultural" debates about the primitive side by side.

Before finally getting to this subject there is one more general point that I want to make about the concept of primitivism. It is one that I feel important to make both because it has not—to the best of my knowledge—been clearly stated before and because it is central to the way in which I shall be discussing primitivism. This is that the term is not only a relational one. It is also a *negative* one. I do not mean this in the sense that it is always used derogatorily, but in the sense that it is defined in terms of what the perceiving culture is *not*. It is the *difference* between the two cultures that causes the sense of the primitive to emerge. It is an absence, rather than a presence, that calls it into being. This fact helps one to understand why such disparate artifacts can be heaped together under the same heading. What possible connection can there be, after all, between a mask created at some unspecified date in Gabon in western Africa, an altarpiece painted in Italy in the fifteenth century, and a woodcut made by a Breton peasant in France in the nineteenth century? The answer must surely be that each one of them is sufficiently different from the high European culture of that period to be perceived by it as "primitive." This also helps to explain why the concept of the "primitive" is so volatile. For it is dependent upon the values current in the defining culture. In 1850, Italian fifteenth-century painting could be seen as "primitive." Twenty years later—with the advent of the aesthetic movement—it was seen as embodying the height of pictorial and intellectual sophistication.

Seen from this point of view, the "primitive" becomes a kind of reverse image of the values prevalent in a society. In the history of art over the last two centuries it becomes a repository for the alternative to the established convention. As such it has played a major role in furnishing the vocabulary for an avant-garde dedicated to criticism and change, to challenging over-sophistication with the image of nature and the established convention of imitation in painting with the promotion of "abstract" or non-representational elements.

These last two remarks will, I hope, have made a paradox clear that has a particular relevance for the visual arts. The adulation of the "natural" man might seem to fall in with the promotion of a "naturalist" style of painting. Yet "naturalist" painting is already practiced by the high European culture and cannot therefore be perceived as "primitive." The adulation of "natural" man brings into question the status of modes of art that go against the European concept of naturalistic representation. As we all know, there have been many attempts to reconcile the conflict by establishing such art forms as superior types of naturalism. This can be seen, for example, in the claims that medieval art is more "natural" than that of the Renaissance; that "natural" artists such as Constable and Courbet derived their forms in a direct and unmediated way from nature; that impressionists truthfully recorded the "primal" vision; even—perhaps most bizarrely of all—that Cubism offered a more accurate way of describing reality than did academic art. It is the attempts to resolve this paradox as it emerges in mid-nineteenth century Britain that will be the focus of the rest of this chapter.

From Universal Primitivism to the Theory of Degeneration

SINCE I AM COMMITTED to the melding of the chronological and cultural dimensions of the primitive, I must go back now briefly to the mid-eighteenth century. For that was the point at which the concept of the primitive began to take on a new significance in western society. The change is symbolized by the emergence to prominence of Jean Jacques Rousseau, the philosopher who promoted, in his *Discourse on the Origins of Inequality* (1755), the notion of the "noble savage," of mankind in his natural state living with a purity and morality unknown to corrupt modern society.[6] The central role of Rousseau's assertions in the movement to create a more egalitarian society is too large and too well known a phenomenon to be entered into here. Suffice it to say that Rousseau gave the notion of the primitive a topicality, making it seem relevant to the growing critical mood concerning social values. It is instructive from this point of view to see how it was enlisted to support a reformist zeal in the visual arts. When the German art critic Winckelmann sought to exhort his contemporaries to return to an emulation of the purity of ancient Greek art he did so by suggesting that the art of that period was based on the study of a race that had the properties of the modern primitive. In his most influential book, *Thoughts on the Imitation of the Greeks in Painting and Sculpture* (1755), he celebrated the classical simplicity of works such as the famous Apollo Belvedere (fig.3.2) in terms of such a comparison, invoking the example of the native American Indian: "Behold the swift Indian outstripping in pursuit the hare: how briskly his juices circulate! how flexible, how elastic his nerves and muscles! how easy his whole frame! Thus Homer draws his heroes."[7]

Winckelmann had in fact never left Europe and had no experience whatsoever of American Indians. But perhaps it would have made little difference if he had. For those who did travel abroad were equally capable of using the "negative" definition of primitivism to incorporate modern non-European and ancient Greek. In his pioneering work *European Vision and the South Pacific*[8] the distinguished Australian art historian Bernard Smith made us aware a generation ago of the propensity of Europeans to cast the non-European peoples they encountered into the roles of ancient Greeks in the decades following the publications of Rousseau and Winckelmann. Often members of tribes in the Americas and the South Sea Islands were given such names as Agamemnon, Hercules and

Aphrodite. When the painter William Hodges accompanied Captain Cook on his second voyage to Tahiti in 1775 he shared in such enthusiasms. His view of a harbor on Tahiti, painted on his return, cast the country as a classical dream. The landscape is staged like a scene by Claude, and the natives in the foreground display their bodies in an easeful manner reminiscent of the ancient inhabitants of Arcadia.

It is noticeable that in these representations there is no reference to the art of the native American Indians, South Sea Islanders, or other groups who were seen to embody the Rousseauian notion of the "noble savage." Lip service, it was true, was sometimes paid during this period to the "vigor" of such work, and it did turn up as a feature in some topographical prints. But it was never promoted as a model for European artists, and the problem of how the "noble savage" could produce such seemingly "unnatural" work was avoided.

This issue only came into prominence when the image of the "noble savage" began to tarnish towards the end of the eighteenth century, and a different, far less flattering interpretation of the "primitives" of the non-European world

Figure 3.2. *Apollo Belvedere*, ca. 350–320 B.C.E., Roman copy, marble. Vatican Museum, Rome

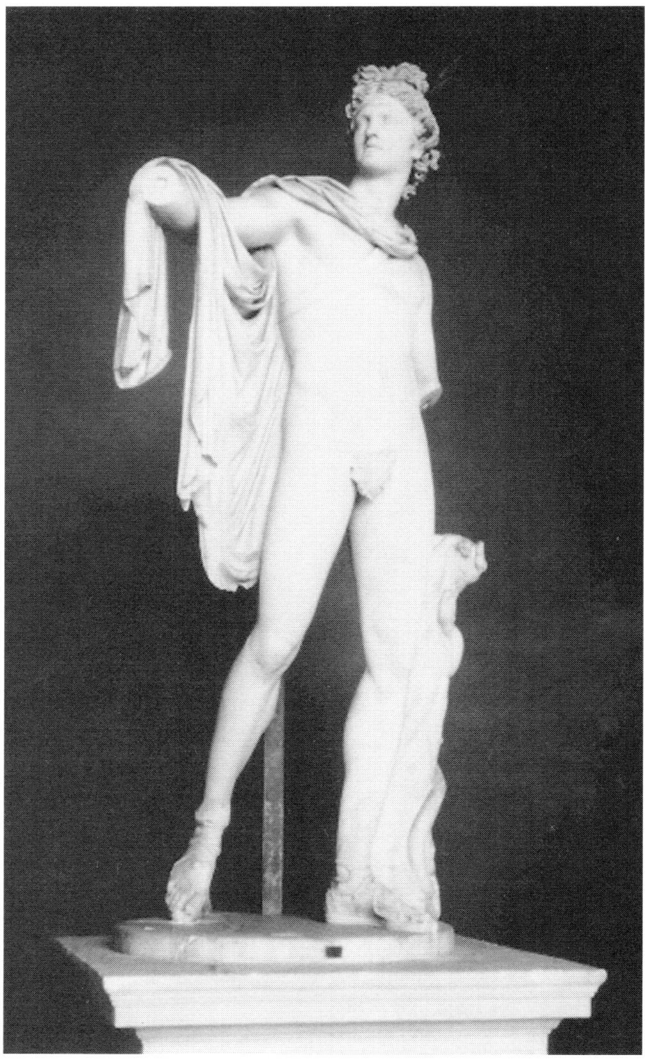

began to emerge. This was what Smith called the "ignoble savage" who was living in a state of brutality rather than primal bliss. To some extent this contrary image was a reaction to the celebratory one, when western observers found, not surprisingly, that the codes of South Sea Islanders were not the same as those of Agamemnon; the assassination of Captain Cook by Maoris was a significant turning point in this development. But far more important for the change of mood were developments within western society. The trauma of the French Revolution and its ramifications in the Napoleonic Wars cast a cloud over the universalism of the eighteenth century and promoted a return to traditional political and religious values. This was not just a revival, however, for it was united with a new and more dynamic concept of evolution that came out of recent scientific developments. The two disciplines combined to form a new vision of man—one that is ably described by Marvin Harris in *The Rise of Anthropological Theory*.[9] In this book, Harris draws attention to the way in which the new concept of evolutionary development became aligned with the biblical account of the Garden of Eden. After being driven from the paradise of Eden for their sins, the parents of mankind had to toil for their living. But there were two paths open to them. One led downward, to an ever lower state of being. The other lay upwards, where by dint of hard work man could begin the long wearisome struggle back to perfection. It was this path of upward toil, it was argued, that westerners had taken. The modern "primitives," on the other hand, had taken the other.

By the mid-Victorian period, this argument had reached the status of a theory called "degenerationism." It had a principle exponent, William Cook Taylor, whose book *The Natural History of Society* was published in 1840.[10] Cook's "degenerationism" was much supported by missionaries—particularly by Protestants who found in it a further justification of salvation by good works and personal effort. Most importantly, the theory separated the modern "primitive" from the idea of "nature." Adam and Eve, after all, had been "natural"—created as God intended them to be. The modern "primitive," by being a degeneration of the "natural," was a distortion of nature. His model could no longer be taken as a guide. Rather, he must be saved from his degeneracy and helped onto the upward path that western man was taking.

One of the interesting points that emerges from this argument is that "nature" itself remains a guide. This is a critical point to grasp when considering the implications of the "degenerationalist" theory for the visual arts. For it enabled commentators to use the unfamiliar forms of the arts of these people to suggest how their lack of conformity to the conventions of naturalism in western art mirrored a lack of naturalness within the society that produced them.

The argument is an unpleasant one, particularly when it occurred in its crudest form, in the journalism of the day and in religious and political tracts. An example of the former can be found in the *Illustrated London News* in 1843 (fig. 3.3), which shows natives of the Marquesas Islands caricatured with their tribal totems.[11] The text makes explicit that the appearance of the statues indicate the low state of the society that they represent. A direct attack on them is deflected by the main point of the news item, which is that these people had recently become subjects (or "allies" as the text puts it) of the traditional British adversary, the French. No doubt the illustrator wished to suggest, by showing one islander holding up a rat as an example of a local delicacy, an affinity between them and the French, with their "unnatural" custom of devouring snails and frogs legs. The savageness of the French natives is brought out by a further contrast. The article on the Marquesas Islands was preceded by one on the British-owned Sandwich Islands. The inhabitants here were praised for their natural

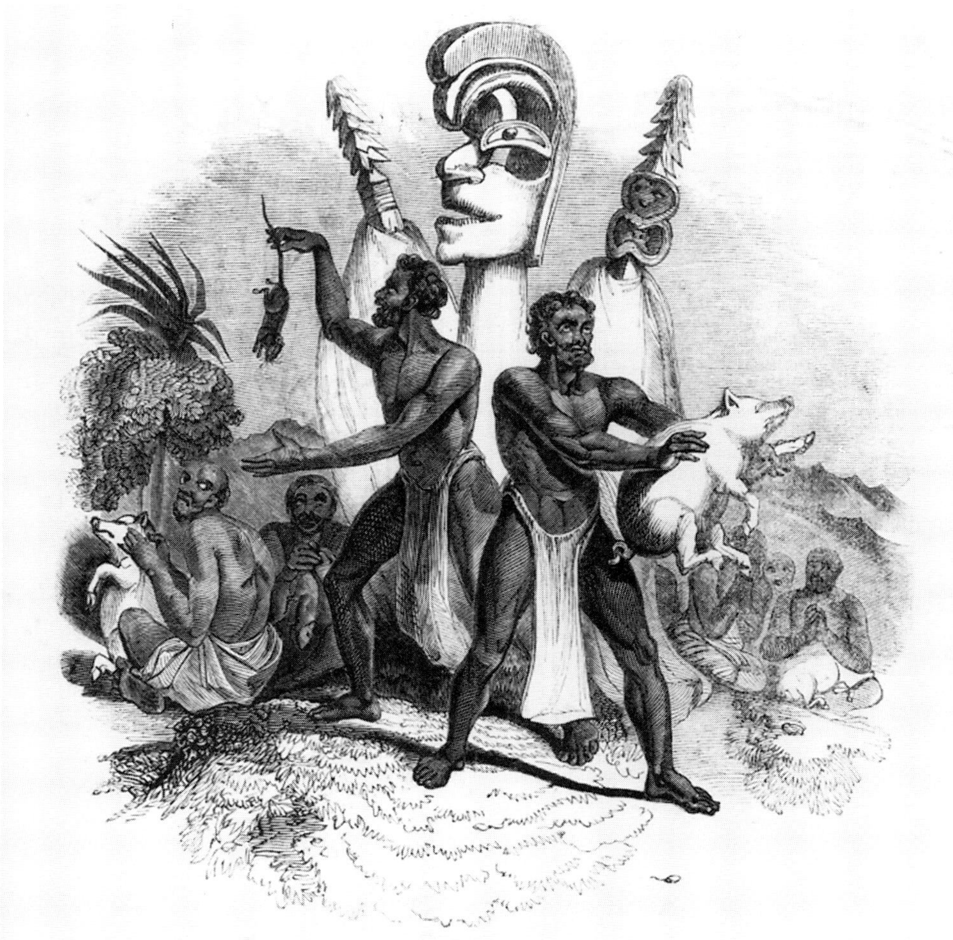

THE MARQUESAS ISLANDERS.

Figure 3.3. "The Marquesas Islanders," wood engraving, *Illustrated London News*, Feb. 4, 1843.

dignity and were represented in robes resembling those of a Roman senator (fig. 3.4). Interestingly, there was no image of the art of the Sandwich Islands included, since this might have undermined such an interpretation. The visual arts of "primitive" societies were only used as evidence of degeneracy at this time and never as an indication of positive virtues.

Another example is of a far more pernicious kind. It comes from the 1852 volume of the periodical *Missionary Scenes* (fig. 3.5) and is commented on by Bernard Smith in the work cited above.[12] The picture shows a New Zealand battleship. In this case the caricature of the tribal art on the prow of the boat is of the crudest kind—as can be seen by comparing it to the carving on the prow of an actual Maori warship. Smith relates the visual interpretation given here to the Maori art to verbal characterization of them by missionaries. These suggested that the art of the Maoris reflected the inadequacy of the religion of the Maoris and, by implication, the necessity of the missionary societies to continue their policy of converting natives to Christianity. As one commentary puts it, "what sort of deities must they be, of which images, as ridiculously fantastic, so monstrously uncouth, so frightfully distorted, as many heathen idols are, are yet considered by their worshipers as appropriate and worthy representatives."[13]

These two examples provide an indication of that growing tendency in the nineteenth century to use the artifacts of a culture as a means of commenting upon its nature. It was a habit that had grown slowly in the eighteenth century, and, I am afraid, has to be laid firmly at the door of the nascent discipline of art history. For it was this subject that launched the concept of art forms being representative of the "spirit" of the society that produced them. It was Winckelmann who laid the groundwork for this in his later work, *History of Ancient Art* (1764), in which he accounted for the changing character of ancient Greek art in terms of the social developments in the relevant period. In subsequent decades this model was applied to the history of art at large. Its most notable exposition was by the philosopher Hegel, who integrated such associative thinking into his dialectical concept of historical development and arrived at the concept of the "Zeitgeist" or "spirit of the age" which could be captured through cultural artifacts. By the 1820s such associations could be found at the level of popular histories of art such as J. S. Memes's *History of Sculpture, Painting and Architecture* of 1829.

As with the concept of "primitivism" itself, this process was first applied within the history of western art before becoming associated—in the manner we saw above—with non-European cultures. What is particularly significant for the

Figure 3.4. "Native of the Sandwich Islands," wood engraving, *Illustrated London News*, Feb. 4, 1843.

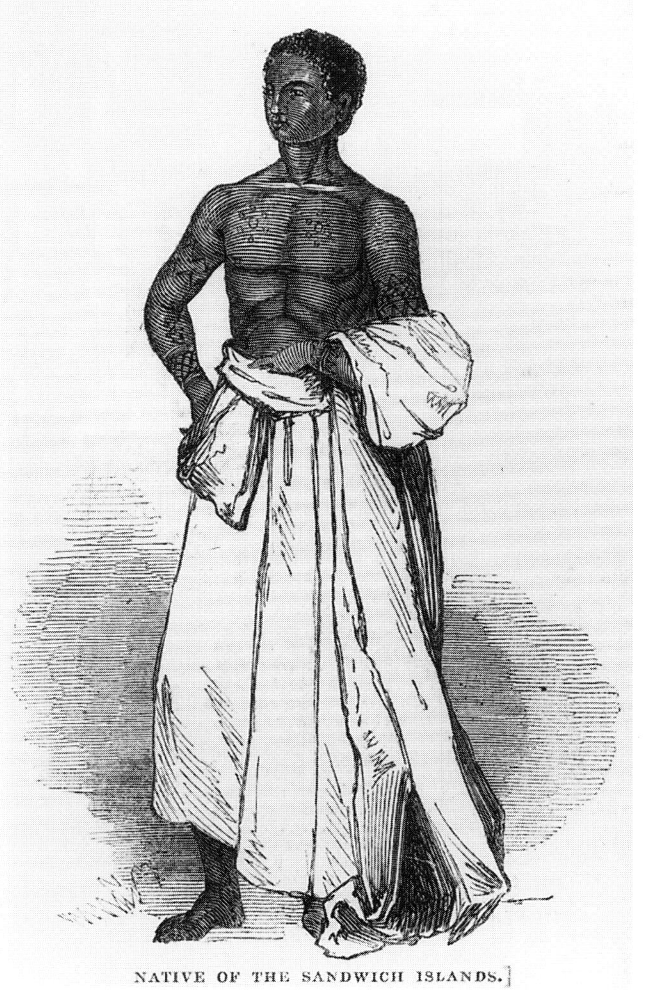

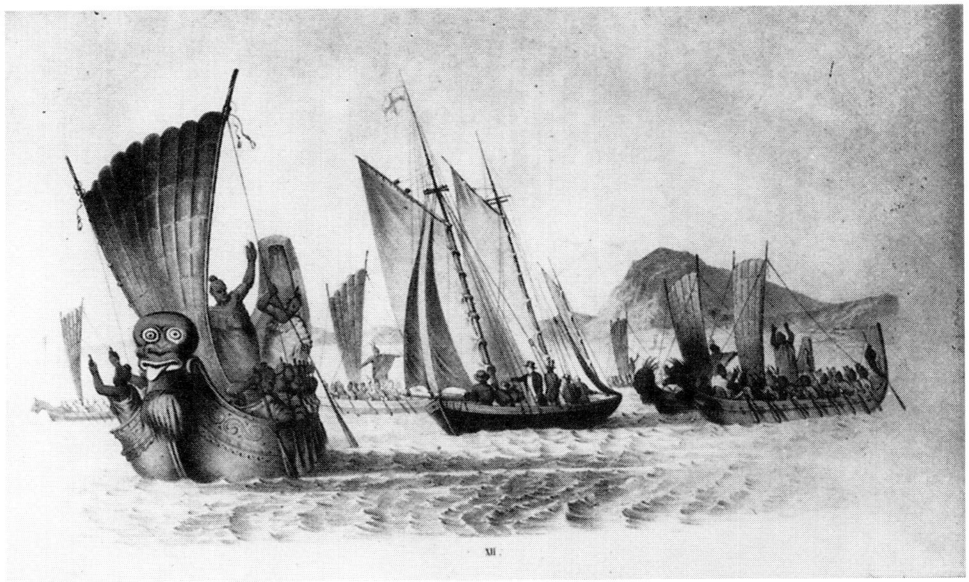

Figure 3.5. "New Zealand War Canoes," engraving, *Missionary Scenes*, vol. 2, no. 12.

theme of this essay is the way in which this combined with the growing nationalism of the period to become a means of establishing the "spiritual" nature of a given country. In the book by Memes cited above the argument was used to associate the peculiar qualities of British art with the country's adherence to the Protestant faith. With this argument the "naturalist" tendency of British art became a symbol of national strength, an indication of the basic independence of mind of the British, a gradual unfolding of the entrepreneurial spirit fueled by the essentially Protestant characteristic of self determination. Memes writes:

> The Reformation, by restoring to the human mind the uncontrolled exercise of its own faculties, by unlocking the barriers by which the will and the powers of free enquiry had been imprisoned, has stamped upon every British institution, as upon every British talent, the worth and manliness of independent character. Our fine arts, though the last to feel, do at length experience this happy influence.[14]

Memes's concern with the institutional roots of culture shares the broad assumptions of the German philosophers of the period, though its precise formulation is more materially based and suggests the inspiration of the Scottish economists and philosophers whose views were prevalent at the time Memes was in Edinburgh.[15]

Memes had in mind as the chief exemplar of this development in the visual arts the aggressive personality of William Hogarth. But even more relevant to his argument was the work of his contemporary David Wilkie, who had gained immense success with his *Chelsea Pensioners Reading the News of the Battle of Waterloo* (fig. 3.6), which shows the populace celebrating the great British triumph, the defeat of Napoleon. Significantly the scene is out of doors—if not quite in the depths of nature, at least with a village atmosphere and a fine summer sky. The "Britishness" of the picture was underlined by the fact that it had been a popular success. It attracted so much attention when it was exhibited at the Royal Academy in 1819 that a bar had to be put round it to hold back the crowds. This was the kind of work that those British Protestant missionaries quoted above would have regarded as "art" and which encouraged them to take

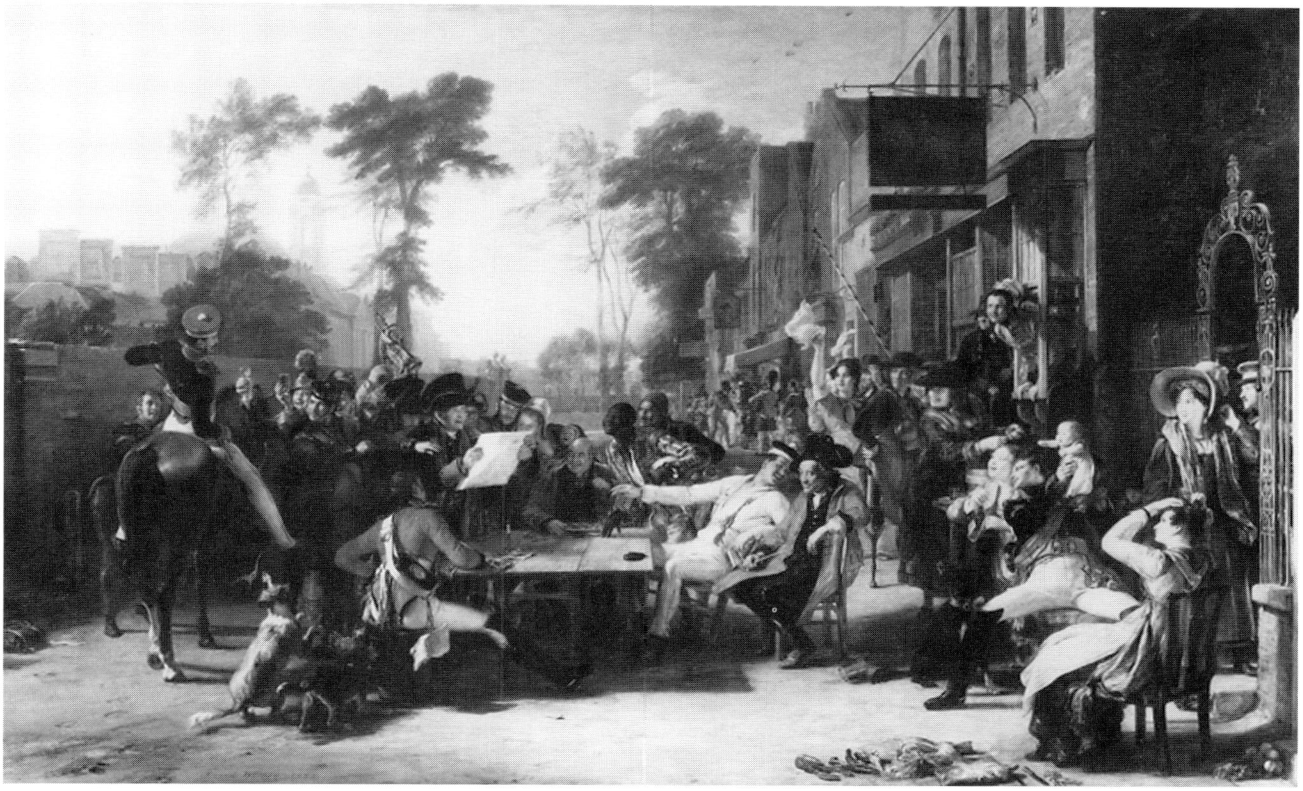

Figure 3.6. Sir David Wilkie (Scotland, 1785–1841), *Chelsea Pensioners Reading the News of the Battle of Waterloo*, 1819, oil. Apsley House, London

such a dim view of Maori carvings. It linked naturalism as closely with the concept of "progress" as it did the so-called "degenerate" and "unnatural" art with the state of the modern primitive.

Abstraction and Revival

THIS IMAGE OF A HEALTHY, nature-inspired art celebrating a progressive nation of self-determinists, however, did not convince everyone. The revived interest in medievalism that ran parallel to the new naturalism had encouraged certain painters to use the forms to explore new expressive and symbolic possibilities. In France Ingres showed in his early career a use of forms derived from both Greek and medieval art that could be dramatically effective in conveying a sense of potency. His portrait of Napoleon in 1806 gives a positively tribal image of the Emperor of the French. Ingres might have been encouraged to use such a strategy by the circumstances of the commission, for this was the time that Napoleon was seeking to invest his government with the trappings of traditional authority. However, this particular mode of primitivism never received official approval, and it subsequently became one of the more "private" features of the artist's style—explored largely for personal aesthetic gratification. But there were others who were returning to a deliberately archaic way of painting for ideological reasons.

Most prominent amongst such groups was the German Nazarenes, who in 1808 sought to return to the supposed purity and spiritual inspiration of the Middle Ages by breaking altogether with the contemporary world. As can be seen in the portrait of one of the group's leaders, Franz Pforr, by another of the leaders, Friedrich Overbeck they grew their hair long and adopted medieval-style clothes while attempting to paint in a manner reminiscent of medieval art. Interestingly, their revolt against the modern world was not treated simply as an absurd eccentricity, but rather was regarded as an uplifting path to be followed and admired, particularly by those that sought to revive the power and authority of the Catholic church.

It might seem that what is emerging here is a division between a largely "Protestant" naturalist tradition and a "Catholic" medieval one. This is indeed very much how the situation came to be read in England—particularly in the 1840s. Yet it is worth remembering that this is not the assumption that was made when the movement was initiated in Germany. In the first place, many of the artists participating in it were in fact Protestants. The "Nazarene" style, it is true, was taken up by the Catholic church in Germany in the post-Napoleonic period, but this institution was far from being its only patron.

As far as the "primitivism" of the works generated is concerned, this never strayed beyond taking models from western Europe. Indeed, an important part of the movement's justification was that it was looking back to a primitive source *within* the European tradition, one that had developed under the specific auspices of Christianity. Furthermore, while attention was paid to the use of certain "abstract" effects for expressive purposes, these were never taken to the point where they actually contravened the naturalistic canon. The Nazarenes themselves made a great point of asserting that their painting—while seeking a simplified effect—did not actually abuse the laws of linear perspective, anatomical proportions, and observation of the fall of light that were the benchmarks of competence for post-Renaissance painters. In fact, strange though this might now sound, they actually defended their art on the grounds that it was more "natural," that is, more honest and direct, than the academic painting they were rejecting. When talking of the work of the early masters that they took for their models, Franz Pforr said, "everything is presented simply, as though it had not been painted but had grown thus."[16] This basic assumption of natural simplicity in the medieval revival was the main topic in Heinrich von Kleist's contemporaneous narrative "On the Marionette Theatre." In this celebrated piece—whose level of irony is difficult to gauge—Kleist tells the story of a dancer who claims that he watches marionettes because of the naturalism of their movements—their simplicity and directness giving gesture and motion in its most perfect and unmediated form. Kleist also mirrored the aspiration to use such contemporary examples of the "primitive" (either in European or non-European societies) as a means of gaining access to a lost primal state of natural being when he wrote, "The doorway to Paradise is locked up, and the Cherub is behind us. We must travel around the World to see if there is anywhere left open at the back."[17]

As I have already suggested, it was not amongst the revivalists themselves, but amongst their opponents that the concept of their art being "unnatural" first arose. And, indeed, it was not until the 1840s that such "unnaturalism" could actually be defended as a virtue. This change of events occurs, I believe, as a reaction to the rampantly "Protestant" characterization of naturalism then prevalent in England. By the 1840s this confident association had begun to look rather tarnished through the dire social consequences of the policy of laissez-faire. English medievalists of this period were happy to question whether the "self-evident" and "liberal" nature of naturalism was really such a good thing.

It was, in fact, not an artist, but a connoisseur who seems to have made the interesting shift to considering that anti-naturalistic elements in the visual arts were worth promoting in their own right. This was the English amateur Lord Lindsay, whose *Sketches of the History of Christian Art* of 1846 was a landmark in the scholarly rediscovery of Italian medieval art. Lindsay, like many English aristocrats, had little time for the entrepreneurial spirit of bourgeois naturalism, and perhaps this is one of the reasons that he was predisposed to take seriously the revivalist pictures of the Nazarenes. But as he immersed himself in the study of the art that the Nazarenes claimed as their source—namely that of the Italian middle ages—he became increasingly convinced that this art was only effective because of disinterest in a naturalistic code. When writing about the frescoes in the Campo Santo at Pisa—that amazing collection of paintings spanning the twelfth century to the fifteenth that were sadly largely destroyed in World War II—he drew particular attention to this point.

When discussing the celebrated last judgment scene—thought then to be by Orcagna but now ascribed to Francesco Traini (fig. 3.7)—he commented that the want of perspective and consistent scale in the picture were not weaknesses in the picture but strengths. Such features, he claimed, "Aid the impression, bring you closer into contact with the poet's thought, ring the passing-bell more loudly in your ear, and place you face to face with domesday."[18]

It might be thought that this was simply a revival of idealistic aesthetics, of the Neoplatonic tradition that had underpinned the arguments for "generalization" in academic art since the Renaissance. But there is a striking difference between this position and that of Lord Lindsay. The "abstract" beauty of the Neoplatonists is one of pure form. Its paradigm is the sphere. It is pure, regular, and—perhaps most important of all—without expression. The "abstraction" of Lord Lindsay—if it might be called such—was expressive. It was an abstraction that intensified the subject rather than attempting to rise above it. As such it seems to prelude the new approach to expressive abstraction that came to fruition in the twentieth century and underlies the link between such an aesthetic and that which heralded a new interest in the primitive.

Within a couple of years of the publication of Lord Lindsay's book there was already a striking development in British art that seemed to show the influence of his approach. This was the establishment in 1848 of the Pre-

Figure 3.7. Francisco Traini (Italy, 1321–1365), engraved by E. Lasinio, *The Last Judgment*, engraving, from *Pitture a fresco del Campo Santo di Pisa* (Pisa, 1812).

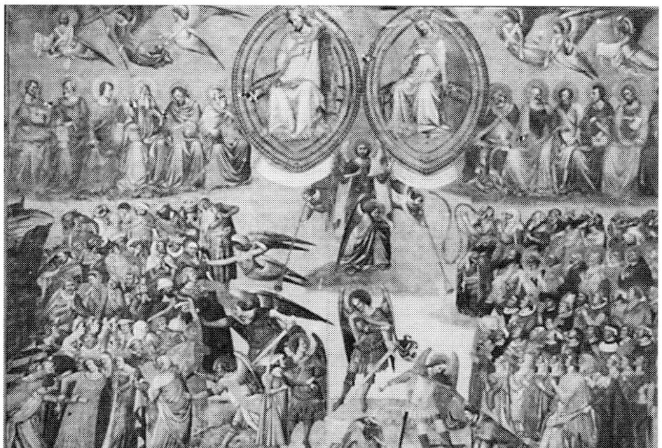

Raphaelite Brotherhood, that group of young students of the Royal Academy in London who, like the German Nazarenes some forty years before them, banded together to break away from current academic practice and to take the art of the early Italians as their model. On the evening in October 1848 when they met together and made this momentous decision they were actually looking through a book of engravings of the Campo Santo frescoes that Lord Lindsay had praised in his *Sketches of the History of Christian Art*.[19] That book also was known to them and may well have impressed them for its comments on the anti-naturalistic basis of early Italian art. Certainly when one looks at some of the Pre-Raphaelites' early work they seem to be going out of their way to use anti-naturalistic elements in a way that the German Nazarenes never did. The painting of the *Ecce Ancilla Domini* [The Annunciation] (fig. 3.8) by Rossetti shows a striking use of spatial distortion and a reduction of the color scheme to a series of blocks of primaries. Nor is this rupture with the naturalistic tradition one simply of slavishly imitating an alternative model. It is combined with a sense of distance, emphasized by the introduction of a contemporary, psychological reading of the story of the Annunciation. The position of the Virgin Mary on the bed and her expression are unprecedented and draw attention to the sexual implications of the story.

Figure 3.8. Dante Gabriel Rossetti (England, 1828–1882), *Ecce Ancilla Domini* [The Annunciation], 1850, oil. The Tate Gallery, London

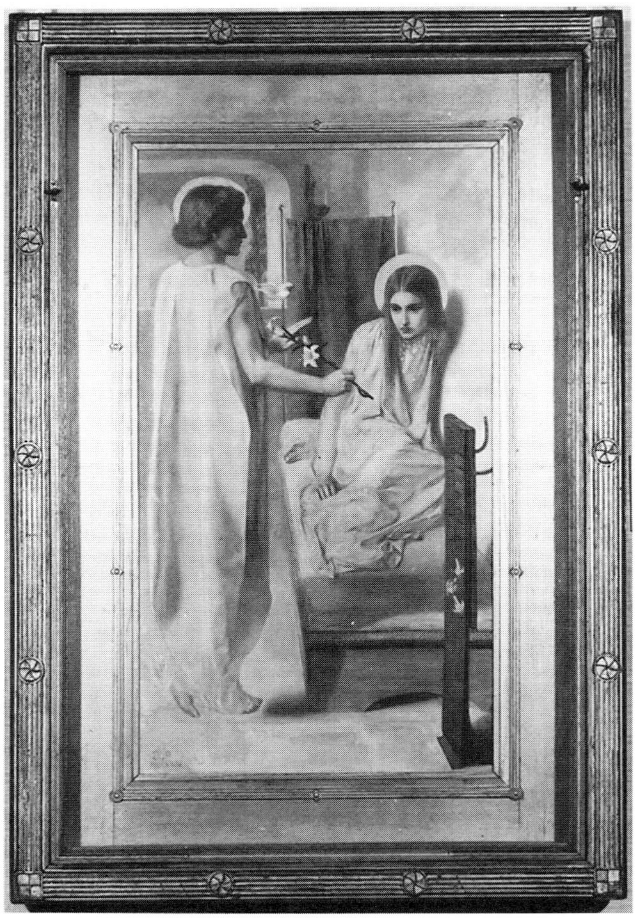

The picture *Christ in the House of His Parents* (fig. 2.8) by Rossetti's fellow Pre-Raphaelite John Everett Millais, also painted in the year 1850, retains a closer link with the naturalistic tradition in its treatment of detail. But it, too, shows a radical use of non-naturalistic elements for expressive purposes. The subject here is an invented one. It shows Christ as a child working in the carpenter's shop of his foster father, Joseph. The child has cut his hand on a nail. His mother, the Virgin Mary, is kneeling beside him in concern. In this trivial childhood incident there is a premonition of his coming crucifixion. Like Rossetti, Millais seems drawn both by what would have seemed to them to be the "quaintness" of medieval art and its expressive potential. He uses an archaic perspectival form, with a high, centralized vanishing point, and pushes his principal figures right to the foreground of the picture, immersing us in their drama in the way that Lindsay had described in the Campo Santo *Last Judgment*. Millais has not actually denied perspectival continuity as Rossetti had done, but he has introduced ambiguities to make it more difficult to read. Look, for example, at the way in which the handle on the vice is set to fall in line with the receding plank on the door laid out on the table. This emphasizes the planar continuity of the picture itself and undermines the illusion of it as a spatial construction. Such usages suggest a sophisticated interest in the non-naturalistic side of medieval primitive art.

As is well known, this moment of rebellion was short lived. Millais' picture was violently attacked by critics when it was exhibited in the Royal Academy. As has long been recognized, the virulence of this onslaught suggests that more than aesthetic niceties were at stake. In her important study *Social and Religious Themes in British Art, 1840–60* Dr. Lindsay Errington has made clear that such opposition was an expression of fears that the "emaciated" and "unnatural" art being produced was an undermining of the British Protestant tradition.[20] Such fears had risen to fever pitch in the very year in which Millais' picture was shown. For the reestablishment of the Catholic hierarchy in England in 1849—something that had been absent from the country for over three hundred years—awakened fears that the British way of life was about to be swamped by an alien spirit. The Pre-Raphaelites themselves were only saved from extinction by the intervention of the critic John Ruskin on their behalf. Ruskin—by far the most influential artistic authority of the day—was able to argue with customary persuasiveness that there was indeed a naturalistic basis to Pre-Raphaelite art. In their later works the Pre-Raphaelites obliged by removing all medieval eccentricities from their style while retaining an interest in a generalized medieval "ethos." They even went so far as to assert that they had never had any interest in what one of them called the "immature perspective, the undeveloped power of drawing, the feebleness of light and shade" of the early Italians.[21]

This story is a sad one and certainly contrasts strongly with the development of an avant-garde in France, where communities of artists would support each other against the rigors of official disapproval. But it is also striking because of the sheer edginess it reveals in the British about their cultural heritage. It is a reminder that, even at the height of their power as an industrial and Imperial nation, the British were perceiving the seeds of their destruction all around. The mid-Victorian period was not one of peace of mind for them. There were already signs of the falling off of the commercial lead they had gained during the Industrial Revolution and of the opposition of other nations within the Empire to their dominance. It is interesting to reflect from this point of view that the depiction of the "degenerate" art of the Maoris in *Missionary Scenes* referred to earlier in this chapter was published at almost exactly the same time that the Pre-Raphaelites were being censured for their lack of naturalism in England.

Within a decade these two strands had been brought together in art debates in London, the heart of the Empire. The international Great Exhibition of 1851, which had been planned to make British industrial preeminence clear to the rest of the world, also brought to the attention of visitors the great wealth of craftsmanship and technical expertise of nations under British rule. They were astonished, for example, by the skill of the textiles of India (fig. 3.9).

Few of the spectators were at that time aware that such magnificent works were the fruits of an industry that was being systematically undermined by the British in India to create a new market for the cheap textile products of Lancashire. But there were those who saw in such works a lesson for British designers who and were anxious to make sure that it would remain before them. It was this that led to the establishment of the South Kensington Museum—now the Victoria and Albert Museum—which set out to gather examples of design from all over the world to provide models for local artists, craftsmen, and designers. It was as a complement to this endeavor that the architect Owen Jones produced in 1857 a classic publication, *Grammar of Ornament*, which provided in

Figure 3.9. "Indian Textile," color lithograph, plate 53 in Owen Jones, *The Grammar of Ornament* (London, 1856).

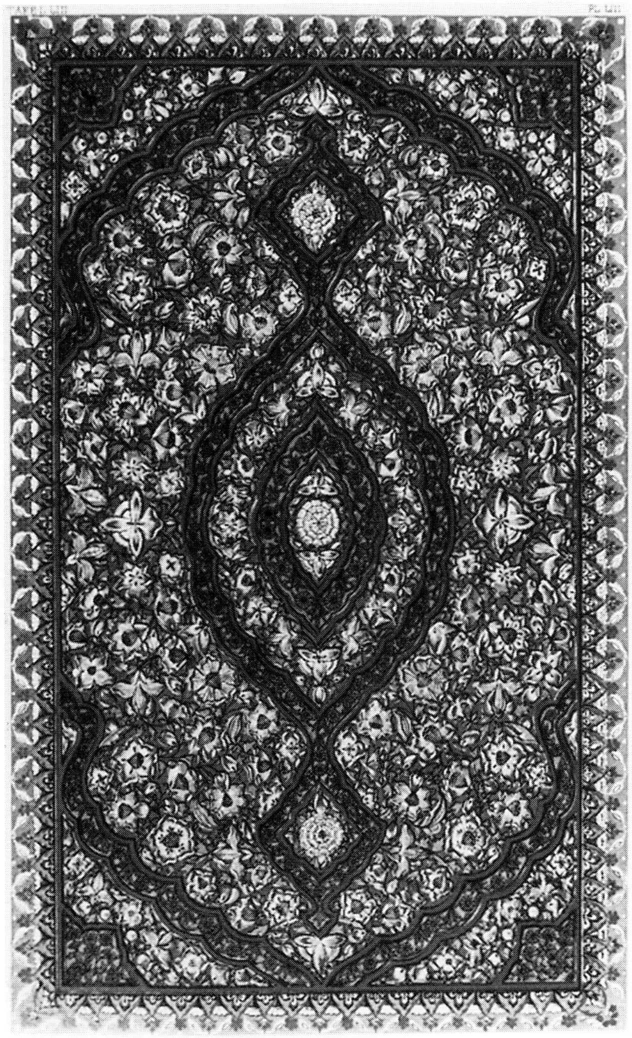

Figure 3.10. "Decorated Maori Head," color lithograph, plate 14 in Owen Jones, *The Grammar of Ornament* (London, 1856).

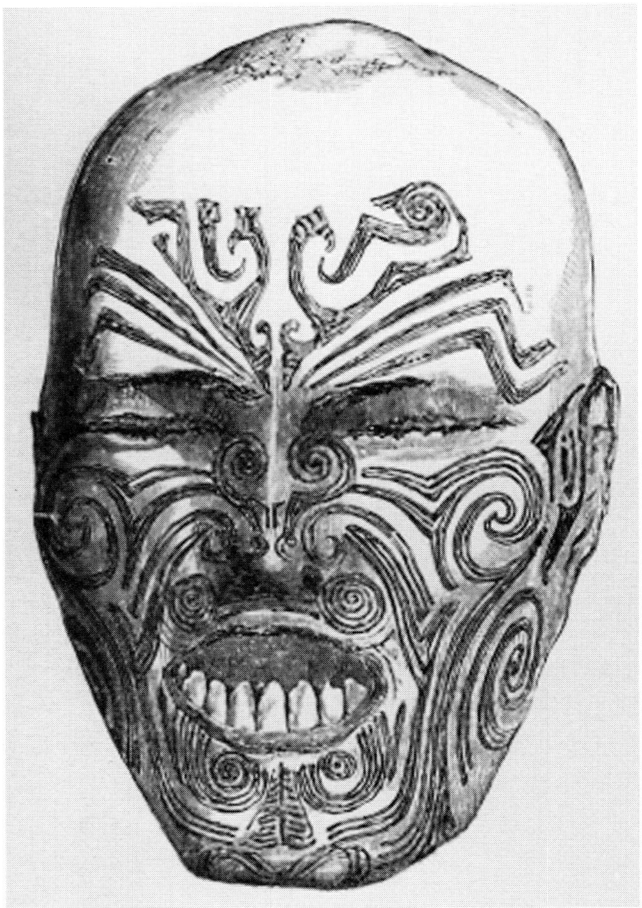

chromolithographs (such as the one reproduced in fig. 3.9) selections for wider distribution and use.

Like Lord Lindsay, Owen Jones provided an alternative voice to that of progressive naturalism. While not denying the importance of nature as a source of inspiration, he asserted that the principles of design were themselves geometrical, rather than imitative. Furthermore, he saw nothing to denigrate in the arts of so-called "primitive" peoples, although the example with which he chose to demonstrate his argument might not have been the one most likely to persuade his audience (fig. 3.10). In his introduction to *The Grammar of Ornament* Jones actually illustrated a decorated Maori head as "evidence of mind," which, he claimed, "will be more readily found in the rude attempts at ornament of a savage tribe than in the innumerable productions of a highly advanced civilization."[22] Jones has, in a sense, rediscovered the Rousseauian noble savage and is using this manifest ingenuity and inventiveness in decoration to resolve the paradox I mentioned earlier. Naturalism is no longer the standard of artistic excellence. It is design.

Jones's publication was one of the initiators of a major reevaluation of the value of design in the visual arts in Europe in the latter part of the nineteenth century. This led not only to a revision of the status of the decorative arts, but also to an increasing concern amongst painters for the role of design and decorative effect in their work. From the start the movement had its adversaries. It is hardly surprising to find that Ruskin was amongst their number. Within a year of the publication of Jones's *Grammar of Ornament* Ruskin had come out with a savage rejoinder. The occasion was an auspicious one, for it was in a lecture that he was invited to give to celebrate the opening of the South Kensington Museum, the institution that had been set up to provide examples of the decorative powers of all nations. Clearly Ruskin felt the need to provide a guide through such work that would counter the influence that Jones's book was already having.

Titled *The Two Paths: The Deteriorative Power of Conventional Art over Nations*, Ruskin's publication sought to reestablish the traditional British view—namely, that art based upon the imitative study of nature indicated a morally healthy society while art based on abstraction, or as he put it "convention," indicated a society in decline.[23] He then goes on to illustrate this in a most unpleasant way. So unpleasant that before repeating it I wish to emphasize that I personally share no part of it.

Ruskin uses as an example the Indian textiles that, as I have already said, astonished visitors to the Great Exhibition by their sophistication and excellence. How is it, Ruskin asks, that this Indian nation—which is known to be degenerate—can produce such work? And how is it that, by contrast, the Scottish highlanders in Britain, known for their stalwartness and moral fiber, have nothing to show of comparable sophistication? His answer is that the Indian art, while elegant, is based on conventional principles. The country has lost its touch with nature and has thereby declined. The Scots, by contrast, though admittedly somewhat lacking in the refinements of high society, still maintain their daily concourse with nature and thereby remain morally pure. If British designers follow the "conventional" example of Indian art then they, too, will be exposing their country to moral danger.

Such an argument is so bizarre that it is tempting to dismiss it out of hand. But one must remember that in 1858 it could gain credence because of one very particular circumstance. At this point in time Britain was still reeling from the effects of the "Indian Mutiny" in which colonialists had been massacred with ghastly brutality. In everyone's minds at the time was the horrendous story of the "Black Hole of Calcutta" in which British women and children were incarcerated in appalling conditions and left to die. These events are commemorated in many works of art of the period, including *In Memoriam* (fig. 3.11) by the Scottish painter Sir Joseph Noel Paton. Originally the picture showed Indian *sepoys* approaching from the rear with murderous intent. But after public protest the artist softened the horror of the event and showed instead the moment of relief—when the survivors were rescued by Scottish soldiers who emerge in the picture's background. This transposition of "bad" Indians for "good" Scots shows how prevalent the polarity was that Ruskin exploited in *The Two Paths*. As Lynn Nead has made clear in her recent book *Myths of Sexuality*, these events had left the British at home in the grips of what has been referred to by social historians as a "moral panic," which spread from the initial shocking events to a kind of witch hunt against all supposed deviant behavior within society and culture.[24] Never had the standard of naturalism seemed more urgently needed.

Ruskin did not confine his attack to non-European decorative art. He carried it into the field of figurative representation within the European tradition. Taking examples from the medieval world, he contrasted the representation of a figure in a tenth-century British illuminated manuscript (fig. 3.12) with a detail from the capital of a twelfth-century church in Milan (fig. 3.13). Though Ruskin's drawing might not make this apparent, the Italian capital shows art in a "healthy" state, already grappling with the problems of naturalist representation. It had long been Ruskin's contention that the Gothic world had shown a nascent naturalism that later triumphed in the art of the Protestant tradition.

Figure 3.11. Sir Joseph Noel Paton (Scotland, 1821–1902), engraved by W. H. Simmons, *In Memoriam*, 1858, engraving.

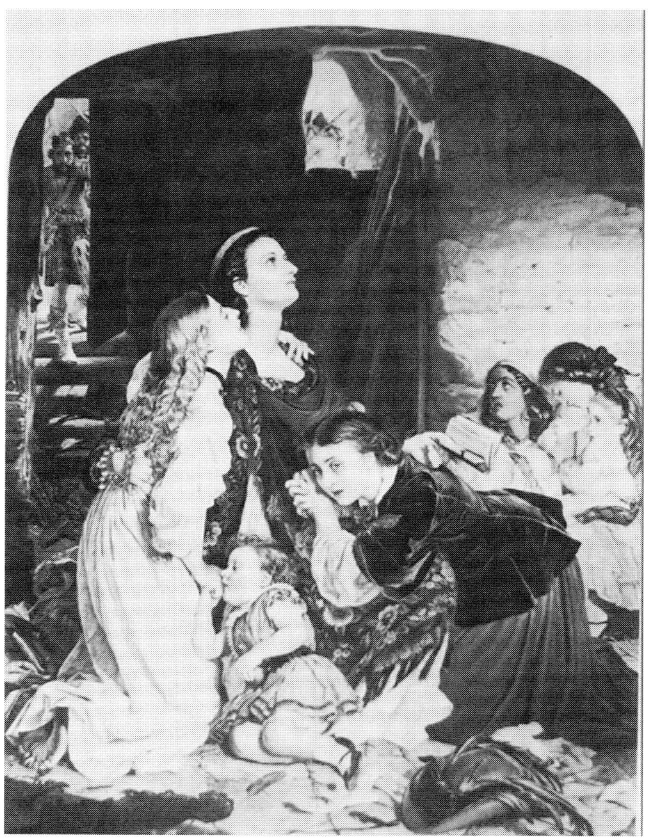

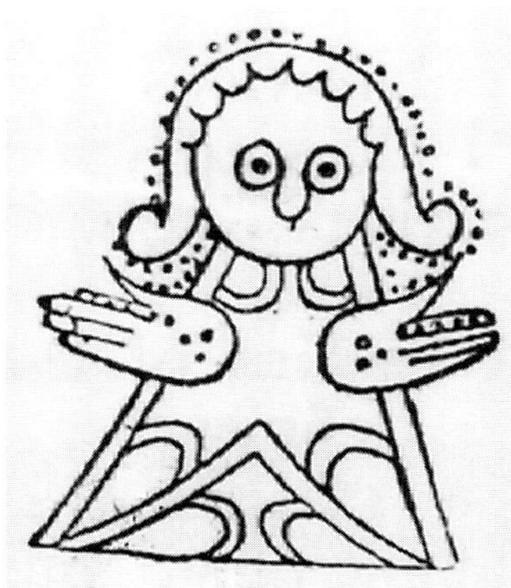

Figure 3.12. Example of "false" figurative art, wood engraving, Ruskin, *The Two Paths*, p. 16 (London, 1859).

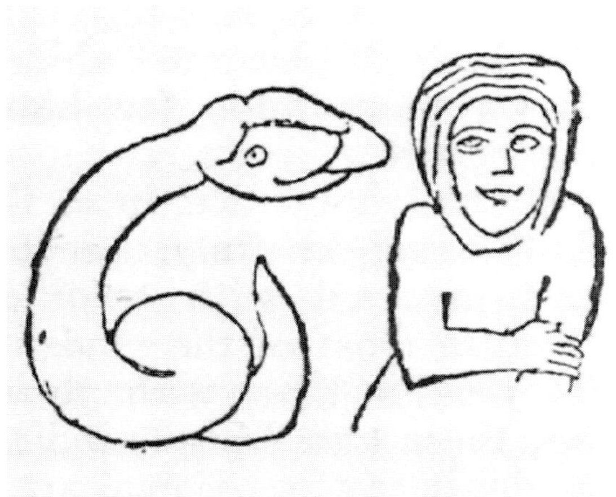

Figure 3.13. Example of "true" figurative art, wood engraving, Ruskin, *The Two Paths*, p. 17 (London, 1859).

What seems curious about this example is that Ruskin should on this occasion choose a British manuscript to illustrate the "degenerate" path. However, Ruskin was extremely knowledgeable on medieval manuscripts, and had chosen his target with precision. His illustration had been taken from a famous publication of the period, Westwood's *Paleographia Sacra Pictoria* of 1843–45 (fig. 3.14). Westwood's publication was a masterpiece of scholarly research. But what worried Ruskin about it was that it concentrated on manuscripts that were from the so-called "dark ages" of early medieval Europe, rather than the Gothic world of the later period, when illuminated manuscripts contained ornament and decoration based on naturalistic details such as the ones that could be readily studied in Owen Jones's *Grammar of Ornament* (fig. 3.15).

Westwood shared Ruskin's disquiet about this particular image. But he offset the shortcomings he found in it against what he described as the "extraordinary skill manifested in the ornamenting of manuscripts" from this period. Not having Ruskin's methodological axe to grind, he saw no problem in praising an art form that demonstrated decorative beauty while lacking in representational attributes. Westwood pointed out—quite correctly—that the decorative work from this generation of British manuscripts became an inspiration for work throughout Europe.

Even so, it might seem strange that Ruskin should have gone out of his way to castigate a British manuscript when he could easily have chosen an example from some other nation. But once again, the target seems to have been chosen with particular precision. For while the manuscript Ruskin chose to attack was British, it was not Anglo-Saxon. It was, in fact, Celtic—either in direct origin or in inspiration. Westwood described it unequivocally as Irish; this was for him a matter of some importance. For one of his objectives in publishing manuscripts from this period had been to show that a great school of Irish decorative art had flourished in the so-called dark ages. In his introduction he claimed to have "clearly established the fact of the existence of a native school of religion and art in our sister island, by a series of facsimiles from Irish manuscripts, executed between the 7th and 10th centuries, although the existence of such documents had been denied, not only by Astle and other Paleographers, but also by the latest Irish historians."[25]

The existence of these superb records of an ancient and highly sophisticat-

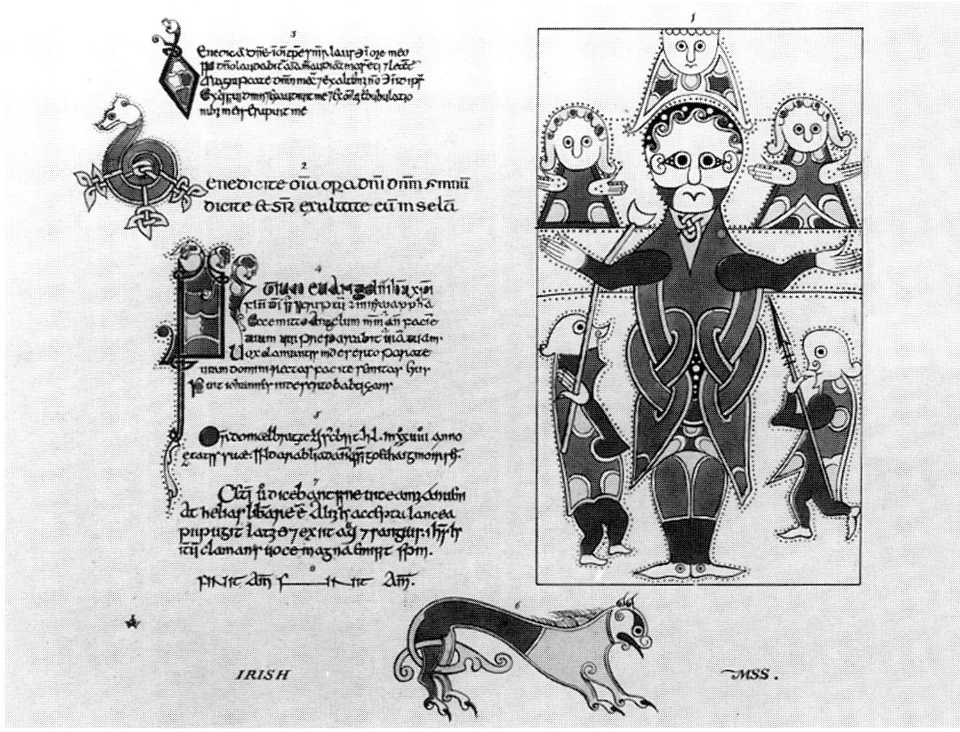

Figure 3.14. "Irish Manuscript," color lithograph, plate 18 in J. O. Westwood, *Paleographia Sacra Pictoria* (London: William Smith, 1843–45).

ed Irish culture may still have been denied by Irish historians in 1843 when Westwood's publications began to appear. But it was hardly so by 1858 when Ruskin delivered his lecture. In the meantime the nascent Irish nationalist movement—spurred on by such tragedies as the potato famine of the 1840s—had begun to take seriously the need to recreate an independent Irish cultural identity. Most attention—not surprisingly—had focused upon the great literary tradition of the Irish, but there had been a growing interest in visual artifacts.[26] A spate of publications came out at this period celebrating the antiquities of Ireland. One such book was Henry O'Neil's *The Most Interesting of the Sculptured High Crosses of Ancient Ireland* (1857) with an engraving showing tourists admiring the High Cross of Monasterboice (fig. 3.16). There were also the first signs of the use of Celtic design for creative purposes—for example, the title page by Margaret Stokes to *The Cromlech on Howth* of 1861 (fig. 3.17) incorporates a capital "T" in the form of a dragon taken from that masterpiece of early Irish illumination, the *Book of Kells*.[27]

Such a development had not gone unnoticed in England. Indeed Owen Jones—himself of Celtic (albeit Welsh) origins—had taken care to include the newly discovered Celtic ornament amongst the examples in his *Grammar of Ornament* (fig. 3.18).

To Englishmen in the grips of a "moral panic" in the late 1850s the idea of Irish cultural predominance—no matter in how distant a period—must have seemed insupportable. Even in the 1840s the concept of Irish culture was seen as comic—as can be seen from a cartoon from *Punch* that shows the Irish leader Daniel O'Connell dressed up in an admittedly somewhat ill-conceived attempt at a national costume. This idea was to become even harder to believe in during the 1860s when there was an upsurge of Fenian activity. As L. P. Curtis has

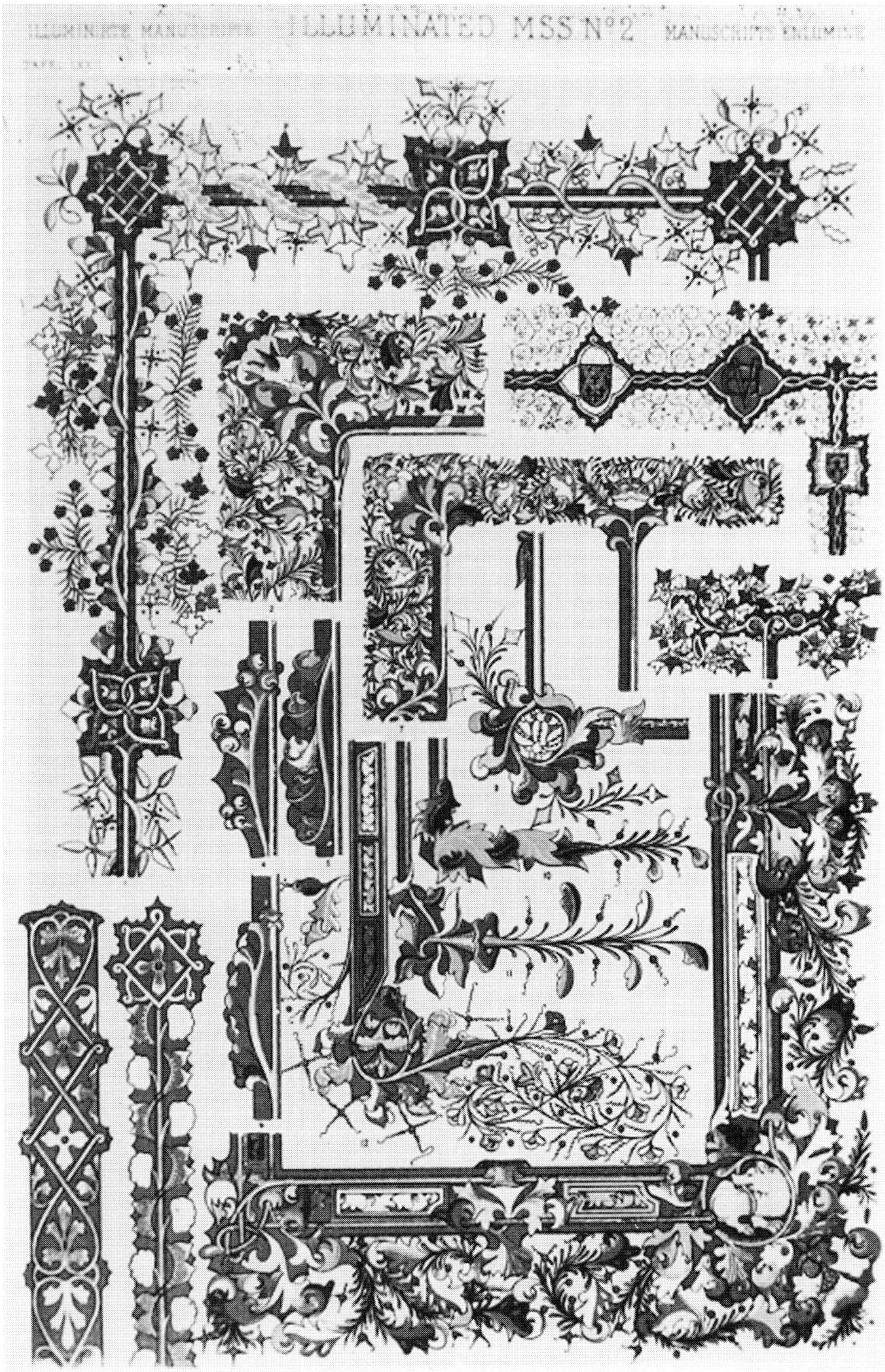

Figure 3.15. "Illuminated Manuscript, No. 2", color lithograph, plate 64 in Owen Jones, *The Grammar of Ornament* (London, 1856).

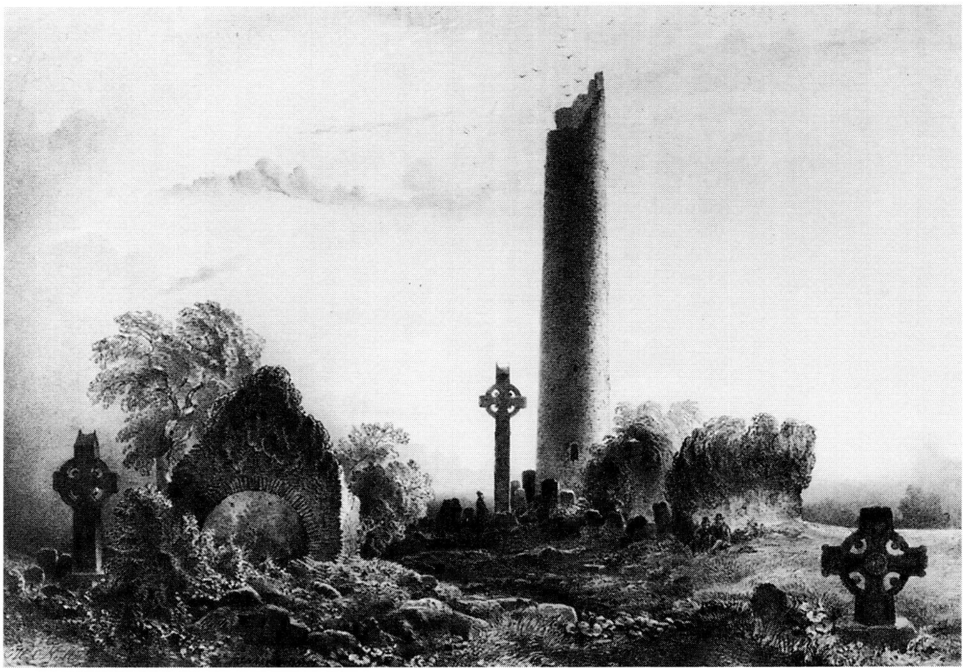

Figure 3.16. *The High Cross of Monasterboice*, engraving, frontispiece to Henry O'Neil, *The Most Interesting of the Sculptured High Crosses of Ancient Ireland* (1857); reproduced in Sheehy, *The Rediscovery of Ireland's Past: The Celtic Revival 1830–1930*, p. 16 (Thames & Hudson, 1980).

shown, this is the period when the distinctiveness and "inferiority" of the indigenous Irish was reinforced by the development by Tenniel and others of a "simianized" Irish stereotype in cartoons.[28] Although the visual arts are not explicitly referred to in such cartoons, it is noticeable that the allegorical image of Ireland that occurs in them (as opposed to the stereotyped image of the indigenous Irishman) is a smaller and weaker version of Britannia. Anything that is civilized about Ireland, this implies, is dependent upon the culture of England.

This is not the place to go any further into the particular history of the Celtic revival. Enough has been said, I hope, to make it clear how Ruskin was able to use the standard of naturalism to isolate the opponents of British Rule. It can serve, too, as a further example of how the identification of primitivism—whether derogatory or adulatory—functions by means of negative definition. In this case it has led to the lumping together of Indians and Irishmen by virtue of the deviations of their separate visual cultures from a supposedly Anglo-Saxon naturalism.

Pictorial Analysis and Racism

In the concluding part of this chapter I would like to consider in brief, almost telescopic form two major developments that seem to me to be affected by this particular Victorian solution to the problem of primitivism and progress, especially by the powerful linking of naturalism with progress and of "convention" (or as some put it "decoration") with stagnation and degeneracy. These developments are the support given to the naturalist convention in the visual arts in the

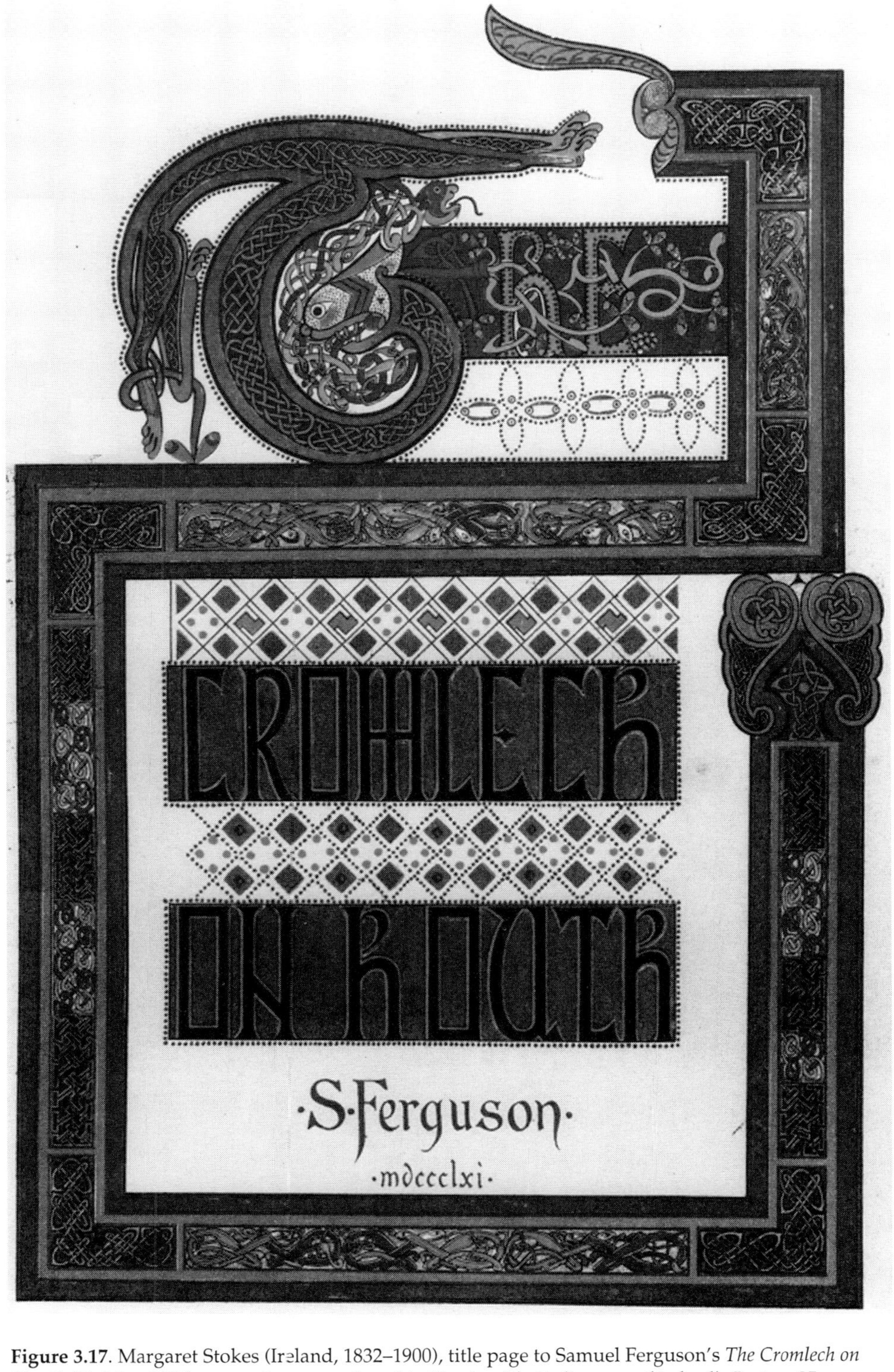

Figure 3.17. Margaret Stokes (Ireland, 1832–1900), title page to Samuel Ferguson's *The Cromlech on Howth* (Dublin, 1861), engraving; reproduced in Sheehy, *The Rediscovery of Ireland's Past*, p. 25.

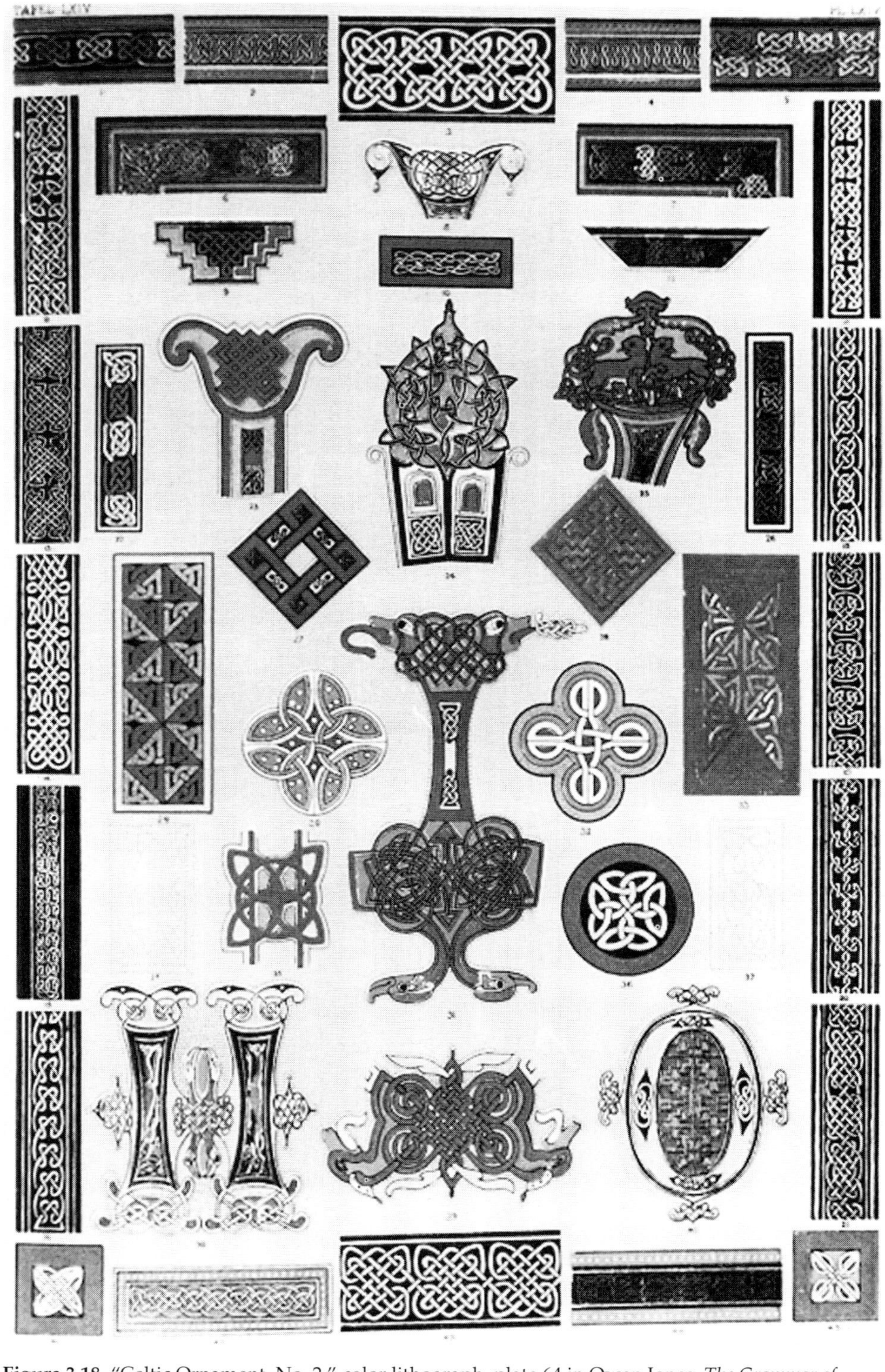

Figure 3.18. "Celtic Ornament, No. 2," color lithograph, plate 64 in Owen Jones, *The Grammar of Ornament* (London, 1856).

very different areas of anthropological theory and the history of British art.

In anthropological theory the ideas of Ruskin certainly contributed to the unfortunate tendency to use degrees of naturalism as a means of discerning "progressive" or "retrogressive" tendencies in cultures. Ruskin would no doubt have been mortified to have found that here his ideas were amalgamated with those of another of his adversaries, Charles Darwin. They are used in tandem, for example, by the celebrated ethnographer Edward Tylor in his *Primitive Culture* of 1871. In this work Tylor sees cultures as engaged in a "survival of the fittest" struggle. Those that had failed in this struggle were, in his opinion, in a terminal state and had to be exposed:

> It is a harsher, and at times even painful, office of ethnography to expose the remains of crude old cultures which have passed into harmful superstition, and to mark these out for destruction. Yet this work, if less genial, is not less urgently needful for the good of mankind. Thus, active at once in aiding progress and removing hindrance, the science of culture is essentially a reformer's science.[29]

Tylor's attention was directed principally to non-European cultures. But there were others who were prepared to use the new science of anthropology to make assumptions about the culture of Britain. Foremost amongst these was John Beddoe, whose *Races of Britain* (1885) provided a record of forty years' painstaking research into the racial mix of Britain and the impact of this mix on society.[30] Beddoe's findings reinforced long-standing assumptions, notably that the Celt was of a fundamentally different ethnological type to the Anglo-Saxon and that the Celt was more evident in the "lower orders" of the social structure.[31] He gave much attention, too, to the fact that the Celt was reemerging in the heartland of England. This fact was potentially disturbing for those who feared both the rise to power of the working classes and the predominance of the English Protestant tradition that had made the Empire. Beddoe does not comment on the artistic traditions of the different races of the British Isles. However, he does characterize the Celtic culture as being essentially "artisan"—something that would fit in with the perception of it as being decorative rather than figurative.[32]

The tendency to make an evaluative relationship between artistic and racial type was taken a stage further in 1893 by Arthur Haddon. In an article in *Illustrated Archaeology*, Haddon made a direct link between the racial purity of a people and the uniformity of their art. "The fluorescence of decorative art," he added "is a frequent result of racial mix."[33] Haddon's view—which was developed further a couple of years later in his book *Evolution in Art*[34]—brings us perilously close to the racism of the twentieth century, particularly in his implied association between racial degeneracy through "mixing" and the decorative tendency. Its relevance for an imperialist country controlling a vast conglomeration of different nations and striving to maintain the dominance of one racial group within it is not hard to find.

It would be unfair to claim that Ruskin had much of a hand in this. He may have expressed the most insupportable prejudices about different cultural groups, but he never claimed—as far as I am aware—that such distinctions were based on anything genetic. Any human, in his eyes, was equally capable of heeding the voice of nature. He attacked certain cultures because he thought they had chosen not to do this, not because he thought they were inherently incapable of doing so.

Such damage as may have been done to the study of anthropology by such ideas has surely by now been rectified. Already in 1911 a new and decisive direction was set by Franz Boas in *The Mind of Primitive Man*. This work conclusively refuted—at the scientific level at least—the notion of there being any connection between racial and mental type and directed attention towards cultural, rather than genetic features as the determinants of a given form of art.[35]

In the case of the interpretation of the history of British art the matter is more complex. It is certainly the case that the distinction between English and Celtic cultures implied in Ruskin's *Two Paths* was increasingly emphasized in the later nineteenth century and encouraged the view that Anglo-Saxons and Celts possessed enduring artistic characteristics that were racially determined. Celticness increasingly came to be seen as something that was "decorative" and out of the mainstream of British art. English art, by contrast, had an innate naturalistic tendency. Even when engaged in decoration, the Anglo-Saxon always operated within the restraints of an overriding naturalistic aesthetic. I am particularly struck by this in the work of two of Ruskin's admirers and disciples, the decorative artist and reformer William Morris and the principal sustainer of Pre-Raphaelite art in its later years, Edward Burne-Jones.

Both Morris and Burne-Jones—who indeed were close friends and worked together on many projects—had a strong involvement in the growing sense of the decorative that I mentioned as a developing principle in European art in the later nineteenth century. But neither were able to follow their more avant-garde contemporaries—such as Whistler in England or Monet in France—in pursuing this beyond a certain point. When Whistler took Ruskin to court for the writer's attack on his painting the "falling rocket" as "flinging a pot of paint in the eyes of the public," Burne-Jones was one of those who entered the witness box to say that such an unnaturalistic daub could not be art. Morris defended a similar view in many lectures and publications. His *Hopes and Fears for Art* of 1881, which seems to have been an attack on Whistler's *Ten O'Clock Lecture*, reasserts the need for naturalism and speaks disparagingly of Japanese art for its failure in this direction: its fatal reliance on formalism. Morris's own decorative work was always based upon the forms of nature and imitates the organic flow of vegetation in its forms.

This position is all the more intriguing because both Morris and Burne-Jones were conscious of their Celtic origins. They were proud of their "Welshness," which proved a common bond between them. But they considered the Welsh to be an inartistic people. "Like his countrymen," Burne-Jones remarked of his father, "he knew nothing at all of art."[36] Neither artist used any of the traditional "geometric" Celtic forms in their work—despite the fact that both clearly had a penchant for decoration. This is surprising in the case of Burne-Jones since, in his later years, he actually produced designs for the revived national Eisteddfod. He encouraged, too, a reading of his art in terms of the Celtic tradition. This is reflected, for example, in the comment of one of his near contemporary biographers, Otto von Schleinitz. In 1901 Schleinitz wrote, "The family originated in Wales. Followers of the atavistic doctrine will perceive in the Celtic temperament of Burne-Jones a valuable proof of their theory."[37]

Burne-Jones' strategy of both promoting and identifying himself with a Celtic revival and at the same time condemning the Welsh as inartistic is hard to fathom. It may be that he was intending to provide the "inartistic" country of his origins with an art form that showed a sense of nature lacking in its indigenous products. He could, in fact, have been indulging in a bit of cultural imperialism. Yet on the other hand, the sheer lethargy and sense of defeat evident in his art and life make other interpretations possible. The "twilight" vision of Celticness certainly appealed to him. He was ever willing to turn his back on the present. He did not seek to revive and reform as did his friend William Morris. One can imagine him relishing the idea of being a member of a conquered race and producing forms of art whose inherent tendency towards decorativeness proclaimed their lack of future.

The impact of such thinking upon those in Wales, Scotland and Ireland who were concerned with reviving "Celtic" forms of art in the late nineteenth and early twentieth centuries is a rich theme that cannot be pursued any further here. It has to be considered in conjunction with the continued concern of English artists—or at any rate of artists working in London and other English cultural centers—with the maintenance of a vigorous naturalist tradition. It is striking how even such avant-garde groups as the New English Art Club were constantly limiting their visual experimentation when it strayed too far away from an English naturalist tradition. The freezing out of the "Glasgow Boys" from their exhibitions in the 1890s is an interesting instance of this.[38] For the Scottish artists (to some extent under the influence of the Celtic revivalists) were veering far more radically towards decoration and abstraction. It does indeed seem that the cultural identification of Englishness with naturalism limited the general range of action of artists in the British Isles in the later nineteenth century and certainly had damaging effects on experimentation in art in the early twentieth century.

Such thinking has also permeated much of the writing of the history of British art, particularly in those books that deliberately set out to provide a general characterization. Here one must acknowledge the impact on the debate about style and national character of the works of the great historian of art Heinrich Woelfflin. Woelfflin belonged to the same generation as those anthropologists who saw a connection between artistic production and racial type. In his *Principles of Art History* of 1915 he makes it clear that he regards race as one of the ineradicable characteristics of style. As he says in his introduction, "to the personal style must be added *the style of the school, the country, the race*."[39] I am not one of those who would try to argue that Woelfflin's inclusion of a racial element seriously undermines the interest of his general theory, though I do find it extraordinary that his remarks about race can occur in books currently available in paperback without any form of editorial gloss or disclaimer. Since Woelfflin's principal writings are concerned with distinctions of period, the racial element in his characterization of style does not predominate. It is a different matter when the Woelfflinian technique is applied to historical surveys of the art of a particular country. For in these circumstances the racial element emerges almost inevitably as the constant. Periods, individuals, political constitutions come and go. The racial element remains.

The most famous treatment of the subject, Sir Nikolaus Pevsner's *The Englishness of English Art*, certainly bears this out. It would be utterly wrong to imply any conscious intention of racial discrimination in Pevsner's work. His main concern was to give a positive aesthetic characterization for an art form that he felt had been undervalued. But in his analysis he built unwittingly on existing prejudice. The very use of "Englishness" as a synonym for "Britishness" is an indication of this. "Englishness" in Pevsner's eyes is rooted in an innate Anglo-Saxon naturalness. "Celticness" is also a tangible factor in the book; characteristically, however, it is given a very minor part in the development of English art. In the whole of his book Pevsner devotes no more than half a page to the issue of Celticness—though he does muse interestingly on the thought that it was Celtic tradition that inspired Britain's most notable contribution to the modern movement, the architecture and design of the Scot Charles Rennie Mackintosh. Pevsner seems at the same time to be accepting the Ruskinian argument that "progressive" English art is essentially naturalistic and to be pointing to the way that belief in this tradition limited the scope of British art in the early twentieth century.

When I began working on the theme of this chapter, I envisaged concluding by emphasizing two points. First, that certain assumptions about the primitive—fueled by colonial experience—had given English artists a peculiarly strong vested interest in the naturalistic tradition in the nineteenth and twentieth centuries. Second, that this concern for naturalism had led historians of British art to marginalize the "non-English" cultures of the country and thereby to distort the dynamics of the artistic tradition. It had been my intention to call for a reevaluation of the history of art in Britain, giving more attention to the tensions that have existed between different cultural groups.

It did not occur to me that there might be anything worth commenting on about the production and evaluation of contemporary art. However, recent events have changed this. In particular, the new criticism of modernism has undermined the valuation of "primitive" abstraction that had been such a problem for British traditionalists in the earlier twentieth century. This has led to a reevaluation of figurative and representational art. While no bad thing in itself, such a reevaluation runs great risks if it is undertaken without critical rigor and a sense of historical development. One of the greatest dangers is that of unreflectively accepting imperialist interpretations of naturalism from the nineteenth century. The recently established magazine *Modern Painters* has revived the claims for British representational art in a manner that intentionally mimics the chauvinism evident in Ruskin in his less perceptive moments.[40] This is a strong reminder that no society can avoid the mistakes and misconceptions of the past simply by ignoring them. For in due course someone will always emerge who is prepared to scour the scrap heap of discarded ideas to find something they can make use of. In the present political climate this particular claim has an ominous ring to it, fitting in as it does with a growing isolationism in Britain and emphasis on "cultural heritage." All the more reason, therefore, to reopen the question of precisely what lies behind the Victorian linking of naturalistic art, Englishness, and progress. For what is at stake is not only how we perceive the past, but also how we understand what is going on around us.

Notes

1. R. Goldwater, *Primitivism in Modern Art* (Cambridge, Mass: Belkamp Press, 1986); and W. Rubin, *"Primitivism" in 20th Century Art* (New York: Museum of Modern Art, 1984).

2. N. H. Graburn, ed., *Ethnic and Tourist Arts* (Berkeley: University of California Press, 1976), 3.

3. A. O. Lovejoy and G. Boas, *Primitivism and Related Ideas in Antiquity* (Baltimore: Johns Hopkins Press, 1935), 1.

4. See discussion at fig. 2.10, p. 46.

5. E. Said, *Orientalism* (New York: Pantheon Books, 1978).

6. For a recent account of these issues see A. Horowitz, "Rousseau's Historical Anthropology," *Rousseau, Nature, History* (Toronto, Ont: University of Toronto, 1987), 50–85.

7. Translated by H. Fuseli, 1766. Reprinted in D. Irwin, *Winckelmann Writings on Art* (London: Phaidon, 1972), 62.

8. B. Smith, *European Vision and the South Pacific* (Oxford: Clarendon Press, 1960; rev. ed. New Haven,Conn., 1985).

9. M. Harris, *The Rise of Anthropological Theory* (New York: Crowell, 1968), 58ff.

10. W. C. Taylor, *The Natural History of Society* (London: Longman, 1840) 1:309–10.

11. *Illustrated London News*, 2 (1843): 68.

12. Smith, *European Vision and the South Pacific*, 244.

13. From *Missionary Sketches*, no. 3 (October 1818). Quoted in Smith, ibid, 245.

14. J. S. Memes, *A History of Sculpture, Painting and Architecture* (Edinburgh, Scotland: Constable, 1829), 208.

15. For an account of the "Scottish Enlightenment" views on art and society see A. Hemingway, "The Sociology of Taste in the Scottish Enlightenment," *The Oxford Art Journal* 12, no. 2 (1989): 3–35.

16. "Einfach stand alles da, als waere es nicht gemalt sondern so gewachsen," Franz Pforr. See M. Howitt, *Friedrich Overbeck* ([Photomechan.] Nach druck der Ausg. der Herder'shen Verlagsbuchhandlung. Freiburg,1886) (Reproduced in Bern: Herbert Land, 1971.) 1:83.

17. "Doch das Paradies ist verriegelt und der Cherub hinter uns; wir muessen die Reise um die Welt machen und sehen, ob es vielleicht von hinten irgendwo wieder offen ist." Heinrich von Kleist "Ueber das Marionettentheater," *Saemtliche Werke* (Leipzig, Germany: Insel Verlag, 1882), 1138.

18. Lord Lindsay, *Sketches of the History of Christian Art*, (London: John Murray, 1846–47), 3:139. This passage is quoted and commented upon by R. Cooper in "'The Crowning Glory of Pisa': Nineteenth Century Reactions to the Campo Santo," *Italian Studies* 37 (1982): 72–100.

19. Hunt, *Pre-Raphaelitism*, 130.

20. See chapter two, p. 35 in this manuscript.

21. Hunt, *Pre-Raphaelitism*, 133.

22. O. Jones, *The Grammar of Ornament* (London: Day and Son, 1856), 14.

23. J. Ruskin, *The Two Paths: The Decorative Power of Conventional Art over Nations*, vol. 16 of *The Works of John Ruskin*, ed. E. T. Cook and A. Wedderburn (London: George Allen, 1903–10), 245–425.

24. L. Nead, *Myths of Sexuality* (Oxford: Blackwells, 1988), 81–86.

25. J. O. Westwood, *Paleographia Sacra Pictoria* (London: William Smith, 1843–45), 1.

26. See J. Sheehy, *The Rediscovery of Ireland's Past: The Celtic Revival 1830–1930* (London: Thames & Hudson, 1980); see especially 17–27.

27. Ibid., 27.

28. See Lewis P. Curtis, *Apes and Angels—The Simianization of the Irish* (Newton Abbot: David and Charles, 1971), 29ff.

29. E. Tylor, *Primitive Culture* (London: John Murray, 1871), 2:410.

30. J. Beddoe, *The Races of Great Britain* (Bristol, England: J.W. Arrowsmith, 1885); see also Curtis, 20–21.

31. Beddoe, *Races of Britain*, 136–142.

32. Ibid.

33. A. Haddon, *Illustrated Archaeology* 1(1893): 108.

34. A. Haddon, *Evolution in Art* (London, 1895).

35. F. Boas, *The Mind of Primitive Man* (New York: Macmillan, 1911). See especially 29ff., where Boas states "... the unproved assumption of identity of cultural achievement and of mental ability is founded on an error of judgement."

36. M. Harrison, *Burne-Jones* (London: Barrie and Jenkins, 1989), 1.

37. O. von Schleinitz, *Burne-Jones* (Beilefeld and Leipzig, Germany: Knackfuss, 1901), 3.

38. See R. Billcliffe, *The Glasgow Boys* (London: Phaidon, 1985), 292–300.

39. H. Woelfflin, *Principles of Art History*, trans. M.D. Hottinger (New York: Dover Publications, 1950), 6.

40. P. Fuller, ed., *Modern Painters* (London, 1988).

The Franklin D. Murphy Lectures in Art bring an art historian of international stature to the University of Kansas campus each year. Each scholar conducts a two-week seminar for the Kress Foundation Department of Art History and delivers two major lectures—one at the Spencer Museum of Art and one at the Nelson-Atkins Museum of Art in Kansas City. The lectureship was established by Franklin D. Murphy, chancellor of the University of Kansas from 1951-1960.

Lectures published by the Spencer Museum of Art to date are:

I. Pierre Rosenberg, *Chardin: New Thoughts*, 1983
III. Xia Nai, *Jade and Silk of Han China*, 1983
V. Robert G. Calkins, *Programs of Medieval Illumination*, 1984
VII. Nobuo Tsuji, *Playfulness in Japanese Art*, 1986
VIII. David Rosand, *The Meaning of the Mark: Leonardo and Titian*, 1988
IX. James Cahill, *Three Alternative Histories of Chinese Painting*, 1988
X. William Vaughan, *Art and the Natural World in Nineteenth-Century Britain*, 1990